Nuclear Radiation Detection Materials — 2009

MATERIALS RESEARCH SOCIETY
SYMPOSIUM PROCEEDINGS VOLUME 1164

Nuclear Radiation Detection Materials — 2009

Symposium held April 14–16, 2009, San Francisco, California, U.S.A.

EDITORS:

Michael Fiederle
Albert-Ludwigs-Universitaet Freiburg
Freiburg, Germany

Dale L. Perry
Lawrence Berkeley National Laboratory
Berkeley, California, U.S.A.

Arnold Burger
Fisk University
Nashville, Tennessee, U.S.A.

Larry Franks
Special Technologies Laboratory
Santa Barbara, California, U.S.A.

Kazuhito Yasuda
Nagoya Institute of Technology
Nagoya, Japan

Materials Research Society
Warrendale, Pennsylvania

CAMBRIDGE
UNIVERSITY PRESS

Shaftesbury Road, Cambridge CB2 8EA, United Kingdom

One Liberty Plaza, 20th Floor, New York, NY 10006, USA

477 Williamstown Road, Port Melbourne, VIC 3207, Australia

314–321, 3rd Floor, Plot 3, Splendor Forum, Jasola District Centre, New Delhi – 110025, India

103 Penang Road, #05–06/07, Visioncrest Commercial, Singapore 238467

Cambridge University Press is part of Cambridge University Press & Assessment, a department of the University of Cambridge.

We share the University's mission to contribute to society through the pursuit of education, learning and research at the highest international levels of excellence.

www.cambridge.org
Information on this title: www.cambridge.org/9781605111377

Materials Research Society
506 Keystone Drive, Warrendale, PA 15086
http://www.mrs.org

© Materials Research Society 2010

This publication has been registered with Copyright Clearance Center, Inc. For further information please contact the Copyright Clearance Center, Salem, Massachusetts.

First published 2010
First paperback edition 2012

Single article reprints from this publication are available through University Microfilms Inc., 300 North Zeeb Road, Ann Arbor, MI 48106

A catalogue record for this publication is available from the British Library

CODEN: MRSPDH

ISBN   978-1-605-11137-7   Hardback
ISBN   978-1-107-40828-9   Paperback

CONTENTS

*Invited Paper

NEUTRON DETECTORS

SCINTILLATOR I

*Invited Paper

MATERIALS II

POSTER SESSION

CdTe AND CdZnTe DETECTORS III

*Invited Paper

SCINTILLATOR II

PREFACE

Symposium L, "Nuclear Radiation Detection Materials," held April 14–16 at the 2009 MRS Spring Meeting in San Francisco, California, provides a venue for the presentation of the latest results and discussion of radiation detection materials from both experimental and theoretical standpoints. As advances are made in these areas of materials, additional experimental and theoretical approaches are used to both guide the growth of materials and to characterize the materials that have a wide array of applications for detecting different types of radiation. The types of detector materials include semiconductors and scintillators, which are represented by a variety of binary molecular compounds such as lanthanum halides (LaX_3), zinc oxide (ZnO), and mercuric iodide (HgI_2). Ideally, desired materials used for radiation detection have attributes such as appropriate range band gaps, high atomic numbers of the central element, high densities, perform at room temperature, have strong mechanical properties, and have low cost in terms of their production. There are significant gaps in the knowledge related to these materials that are very important in making radiation detector materials that are of higher quality in terms of their reproducible purity, homogeneity, and mechanical integrity. The topics that are the focal point of this symposium address these issues so that much better detectors may be made in the future.

Michael Fiederle
Dale L. Perry
Arnold Burger
Larry Franks
Kazuhito Yasuda

December 2009

MATERIALS RESEARCH SOCIETY SYMPOSIUM PROCEEDINGS

MATERIALS RESEARCH SOCIETY SYMPOSIUM PROCEEDINGS

Prior Materials Research Society Symposium Proceedings available by contacting Materials Research Society

CdTe and CdZnTe Detectors I

Mater. Res. Soc. Symp. Proc. Vol. 1164 © 2009 Materials Research Society   1164-L03-02

Development of THM Growth Technology
for CdTe Radiation Detectors and the Applications

Minoru Funaki, Hiroyuki Shiraki, Mitsuru Tamaki, Yoshio Mito and Ryoichi Ohno
Acrorad Co., Ltd., 1-27-16 Hamamatsu-cho, Minato-ku, Tokyo, Japan

ABSTRACT

4 Nines (99.99%) Cd and Te were purified to the semiconductor grade 6 Nines ~ 7 Nines purity materials by the distillation and the zone melting processes, in order to be used for the growth of CdTe single crystal. The CdTe single crystal of 100 mm in diameter and 18kg in weight was successfully grown by the traveling heater method (THM). The shape of the growth interface had the key role for the single crystal growth. The distribution of the Te inclusion size was measured by IR microscopy. The uniformity of mobility-lifetime products and energy resolution in the wafer were also evaluated. The CdTe X-ray flat panel detector (FPD) was developed using the THM grown CdTe single crystal wafer. The CdTe pixel detectors with 100 µm pixel pitch were flip-chip bonded with the C-MOS readout ASIC and lined up on the print circuit board to cover the active area of 77 mm x 39 mm. The evaluation results showed that the CdTe X-ray FPD is promising as the imager for the non-destructive testing.

INTRODUCTION

Thanks to the large atomic number (48-52) and the wide band gap (1.5 eV), CdTe was recognized as a prospective material for the room temperature radiation detector in 1970s [1-2]. Since then, various types of growth methods have been attempted, i.e. Te-rich solution growth method, Traveling Heater method (THM), Bridgman method, High Pressure Bridgman method, Vapor Phase growth method and so on.

In case of the THM growth, CdTe crystal is grown from the Te-rich solution at much lower temperature than the melting point (1092°C). Due to this low temperature growth and the impurity gettering effect of the Te-solution, THM has the advantage to grow the higher purity crystal, where the purity is very important for obtaining the larger mobility-lifetime products. On the other hand, it had been believed that "THM can grow only the small diameter crystals (10~20 mm) at very small growth rate, so that it isn't suitable for the production". Although this limit was not based on the theoretical reasons, the improvement of productivity was really crucial as the industrial process.

20 years ago, we started the development of CdTe THM growth technology. The crystal diameter was 32mm and the grown single crystal showed the very good energy resolution and found to be very uniform throughout the crystal [3]. The step by step scale-up of the crystal diameter (32 mm, 50 mm [4] and 75 mm [5]) has been needed with increase in the demand of the large quantity and the low cost in market. Although we could recently produced 700,000 of CdTe detectors in a year by using 75mm diameter crystal, more productivity is still the strong demand in many applications.

Concerning the application of CdTe detectors, the X-ray FPD is an application with interest, as it can be used for the medical instruments, non-destructive testing devices, security, food inspection and so on. For this kind of imaging applications, the wafer size and the uniformity are very important.

In this paper, our recent development of 100 mm diameter CdTe single crystal growth technology by THM and X-ray FPD were discussed.

EXPERIMENT

Purification of raw materials

Commercial 4 Nines grade Cd and Te were purified to the semiconductor grade 6 Nines~7 Nines purity materials by the distillation and the zone melting processes. The distillation was repeated twice for the purification of Cd, the distillation followed by the zone melting were carried out for the purification of Te.

The raw materials were distilled in the high purity graphite containers with evaporation and condensation sections under the high purity hydrogen flow. The distillation yield was kept at about 80%, for both Cd and Te.

The distilled Te was then purified by zone melting in the high purity hydrogen atmosphere. The tail 20% part was removed due to the impurity accumulation by the zone refining process. The material purity was analyzed by glow discharge mass spectrometry (GDMS).

THM growth of 100mm diameter CdTe and the characterization

Prior to the THM process, a CdTe poly-crystal ingot and Te solution alloy were synthesized in the quarts ampoules for the THM growth. For Cl doping of the THM crystal, Te-solutions were doped with appropriated amount of $CdCl_2$. Then the CdTe single crystal seed, the solution alloy and the CdTe poly-crystal ingot were charged into the quarts THM ampoule and the ampoule was evacuated and sealed.

The prepared THM ampoule was set in the THM furnace so that the Te-solution part is located at the middle of the furnace. Then the crystal growth was carried out with raising the furnace temperature and pulling the ampoule downward slowly. The growth rate of 3mm/day~7mm/day was used in this work. The schematic diagram of the THM growth was shown in figure 1.

In order to investigate the growth interface shape during the growth, the furnace was rapidly cooled to room temperature and the growth interface part was cut along the growth axis. For the characterization of Te inclusion, CdTe wafer was observed by infrared (IR) microscopy and the size distribution was studied. The mobility-lifetime ($\mu\tau$) products were also measured using the "$\mu\tau$ model" spectral fitting method [6]. Schottky type detectors [7] with dimension of 4 mm x 4 mm x 1mm were fabricated from the 100 mm diameter wafer and the uniformity of energy resolutions was investigated.

X-ray Flat panel detector

The ohmic type pixel electrodes with 100 µm pitch and the Schottky continuous electrode were fabricated on the each side of the 1 mm thick CdTe single crystal wafer, then the wafer was

4

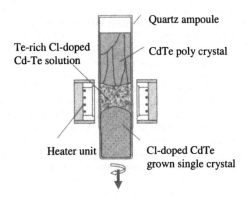

Figure 1. Schematic diagram of CdTe THM growth

Quartz ampoule

Te-rich Cl-doped Cd-Te solution

CdTe poly crystal

Heater unit

Cl-doped CdTe grown single crystal

cut into a couple of individual detector, which size is 12.9 mm x 25.7 mm. The pixel CdTe detector was Flip-Chip bonded with the C-MOS readout ASIC. They were lined up on the print circuit board to form the flat panel with the dimension of 77 mm x 39 mm. The spatial resolution was evaluated by the test chart image. The result was compared with the CsI:Tl flat panel. The detail was described elsewhere [8].

RESULTS and DISCUSSION

The purification of Cd and Te

The picture of purified Cd (distillation twice) and Te (distillation followed by zone refining) was shown in figure 2. The GDMS analytical results of Cd and Te at each purification steps were shown in table I.

Figure 2. The purified Cd (left) by distillation and Te (right) by distillation and zone melting

For Cd purification, the major impurities of raw 4Nine Cd were Cu, Ag, Tl and Pb, with the impurity level of a few ppm in weight base. All these impurity elements are effectively removed by the first distillation and the sum of the detected impurities was 0.059 ppm, resulting that the distilled Cd was 7 Nine grade material. However, still S, Cl, Ti, Cr, Fe, Cu, Pb were detected at ~10 ppb level. These elements could be removed below the detection limit by the second distillation except for Zn, which seems to be harmless element for CdTe radiation detector.

On the other hand, the distillation of 4Nine Te removed Al, Cl, Fe, Co, Ni, Ir, Pb below the detection limits and Cu was reduced by one order, while there was almost no effect on Se. After the distilled Te was purified by zone melting process, only Si and Se were detected. The origin of Si was probably the quarts boat for zone melting. Se was reduced by one third. The sum of the detected impurity was 0.41 ppm.

As described above, the commercial 4N grade Cd and Te were purified to the semiconductor grade materials by coupling distillation and zone refining process.

Table I. GDMS analytical result of each step of the purification process
The heavy element, which was not detected in the raw material, was excluded from the table.

element	GDMS Analytical result (weight ppm)					
	Cd			Te		
	4N Raw material	Distillation once	Distillation 2times	4N Raw material	Distillation	Distillation +Zone melting
Na	0.005	<0.001	<0.001	0.005	0.002	<0.001
Mg	<0.001	<0.001	<0.001	<0.001	<0.001	<0.001
Al	<0.001	<0.001	<0.001	0.03	<0.001	<0.001
Si	0.008	<0.001	0.003	<0.001	0.009	0.01
P	<0.001	<0.001	<0.001	0.006	0.02	<0.001
S	0.38	0.02	<0.005	0.009	0.008	<0.001
Cl	0.003	0.005	<0.001	0.35	<0.01	<0.01
K	<0.005	<0.005	<0.005	<0.05	<0.05	<0.05
Ca	<0.005	<0.005	<0.005	<0.05	<0.01	<0.01
Sc	<0.001	<0.001	<0.001	<0.001	<0.001	<0.001
Ti	<0.001	0.001	<0.001	<0.001	<0.001	<0.001
V	<0.001	<0.001	<0.001	<0.001	<0.001	<0.001
Cr	<0.001	0.003	<0.001	<0.001	<0.001	<0.001
Mn	<0.005	<0.005	<0.005	<0.001	<0.001	<0.001
Fe	0.02	0.02	<0.005	0.006	<0.001	<0.001
Co	0.01	<0.001	<0.001	0.002	<0.001	<0.001
Ni	0.12	<0.001	<0.001	0.39	<0.001	<0.001
Cu	7.2	0.003	<0.001	2.0	0.17	<0.001
Zn	0.02	<0.001	0.01	<0.005	<0.005	<0.005
Se	<0.002	<0.002	<0.002	1.6	1.4	0.4
Ag	4.1	<0.05	<0.05	<0.01	<0.01	<0.01
W	0.01	<0.001	<0.001	<0.001	<0.001	<0.001
Re	0.003	<0.001	<0.001	<0.001	<0.001	<0.001
Ir	<0.001	<0.001	<0.001	0.24	<0.001	<0.001
Hg	0.02	<0.005	<0.005	<0.001	<0.001	<0.001
Tl	1.8	<0.001	<0.001	<0.001	<0.001	<0.001
Pb	3.8	0.007	<0.001	0.02	<0.001	<0.001
Bi	0.01	<0.001	<0.001	<0.001	<0.001	<0.001

6

THM growth of 100 mm diameter CdTe

The picture of the grown CdTe single crystal ingot of 100 mm diameter was shown in figure 3. The total length and the weight were about 390 mm and 18 kg, respectively. We believe that this is the largest THM grown CdTe single crystal that has ever been reported. When we developed CdTe THM growth in 1992, the ingot diameter and weight were only 32mm and 600g, which was also the largest at that time. The crystal weight became 30 times in 17 years, increasing the productivity dramatically.

Figure 3. CdTe ingle crystal ingot grown by THM.

As new grains were generated during the crystal growth at the early stage of this work, the grown crystal became a poly-crystal and the thorough improvement of the growth condition was necessary for the the single crystal growth.

Observing the cross section at growth interface was classical approach but had been the most effective way to improve the single crystal yield through our scale-up experience. The cross sectional view of the growth interface at conventional growth condition was shown in figure 4. As a whole, the growth interface was convex to the solution zone, but a partially concave part existed. Below this point, a newly generated grain boundary was found. This growth interface is the primarily the result of the temperature profile, so that the improvement of thermal condition was carried out to obtain the complete convex shape of the growth interface.

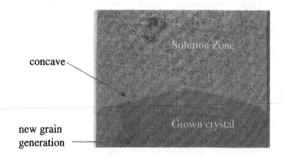

Figure 4. The cross sectional view of the growth interface before optimization of the growth

The effect of growth interface was shown in figure 5. Before the improvement, the growth interface at the initial growth stage included the partial concave part, as already described and seen in right hand side of figure 5. While that at the final growth stage was completely round and convex shape. Probably due to the incomplete convex shape at the initial stage, the grown CdTe crystal includes the grain boundaries close to the wafer edge. On the other hand, the growth interface at the initial growth stage was improved by the change of thermal environmental condition as seen in left side of the figure 5. The interface shape became a complete round and convex shape during the whole growth period. The grown crystal was confirmed to be the single crystal.

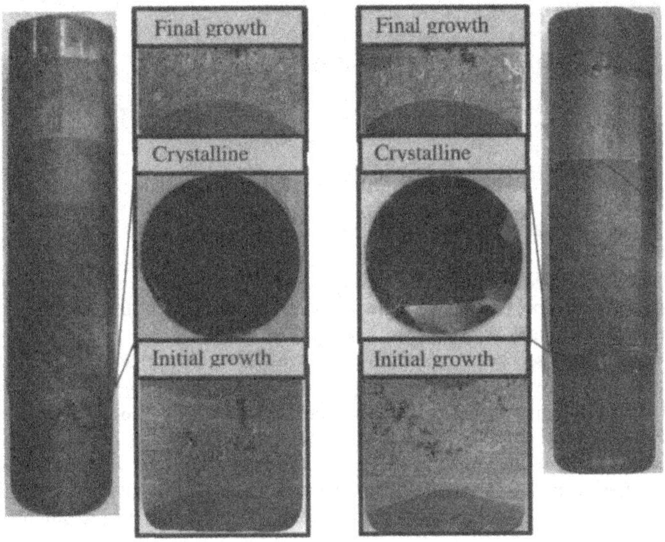

Improved growth condition   Conventional growth condition

Figure 5. The effect of thermal environmental condition improvement on the crystalline quality

Characterization of the grown crystal

The distribution of the Te inclusion size measured by IR microscopy was shown in figure 6. It was found that there were two groups in the distribution. The smaller one was the diameter less than 7 μm and about two third of the Te inclusions were in this group. The larger group was the diameter between 10 μm and 20 μm and the average size was roughly 15 μm. This inclusion size were still acceptable for the X-ray imaging detector with 100 μm pixel pitch, described later.

Concerning the transport properties, the map of the μ-τ products in the 4 inch wafer were shown in figure 7. The average μ-τ products for electrons and holes were 2.5×10^{-3} cm^2/V and 5.0×10^{-4}, respectively. The standard deviations of the μ-τ products both for electrons and holes were less than 10% of the average, showing the very good homogeneity within the wafer.

8

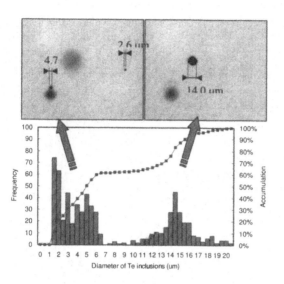

Figure 6. Size distribution and IR image of the Te-inclusion

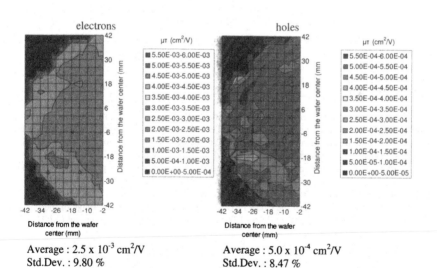

Average : 2.5×10^{-3} cm^2/V
Std.Dev. : 9.80 %

Average : 5.0×10^{-4} cm^2/V
Std.Dev. : 8.47 %

Figure 7. Mapping data of μ-τ products in the wafer

9

The Schottky detectors with dimension of 4 mm x 4 mm x 1 mm were fabricated and the energy spectra of the ^{57}Co were measured. The typical spectrum and the FWHM of the 122 keV peak was mapped in the figure 8. The average of the FWHM and the standard deviation were 5.3keV and about 6% of the average, respectively. This uniformity of the detector performance is very important for the mass production of the detector and for the large area imaging detector.

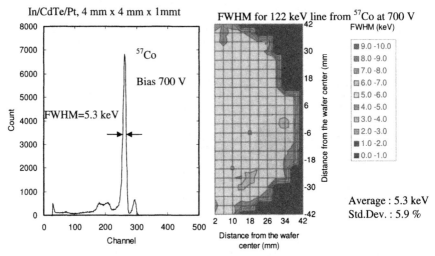

Figure 8. The typical energy spectrum of the Schottky detector (left) and the FWHM mapping in the wafer.

CdTe X-ray flat panel detector

A couple of pixel detectors with the dimension of 12.9 mm x 25.7 mm were simultaneously fabricated on a single crystal CdTe wafer as shown in figure 9. Then the wafer was diced into the individual detector. The CdTe detector was flip-chip bonded with C-MOS readout ASIC, to fabricate the CdTe-ASIC hybrid module, shown in figure 10. Nine pieces of the hybrid were die bonded on the print circuit board to fabricated the flat panel detector with the active area of 39mm x 77 mm as shown in figure 11. The detail of the FPD fabrication was described elsewhere [8].

The spatial resolution of the CdTe FPD was evaluated by taking the image of the resolution chart pattern as shown in figure 12. The image taken by the commercial CsI:Tl flat panel with the same pixel pitch was also shown for comparison. The 5 line pair pattern was clearly seen by the CdTe X-ray FPD. On the other hand, the same pattern taken by the CsI:Tl flat panel was blurred probably due to the diffusion of the emitted light in the CsI:Tl scintillator layer.

In order to assess the potential of the CdTe FPD as the imager for non-destructive testing, many kinds of objects were imaged by the CdTe FPD. Figure 13 shows the image of a USB flash memory as an example. The image correction was performed by adjustment of difference in the gain linearity among the pixels and by replacement of missing data with the average of neighboring pixels. The detail of the flash memory was clearly seen, showing the potential as a new type of FPD.

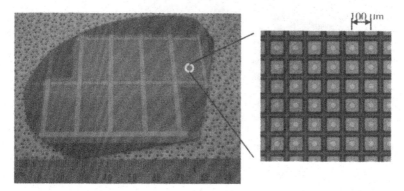

Figure 9. The CdTe single crystal wafer with pixel electrode

Figure 10. CdTe –ASIC hybrid

Figure 11. CdTe X-ray FPD

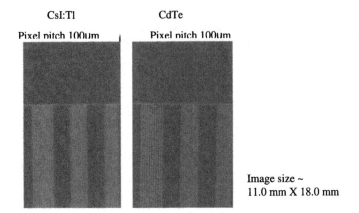

Image size ~
11.0 mm X 18.0 mm

Figure 12. The comparison of X-ray images of a resolution chart taken by the CsI:Tl (left) and the CdTe module at 80 kV.

Figure 13. The image of USB flash memory.

CONCLUSIONS

The 4 Nines grade Cd and Te were purified to semiconductor grade 6 Nines ~ 7 Nines grade purity by the distillation and the zone melting processes in house. The CdTe single crystal of 100 mm diameter was successfully grown by the improvement of the growth conditions. The observation of the Te inclusion revealed that the size of Te inclusion was divided into two groups. Larger one is from 10 to 20 μm in diameter and the smaller one is less than 7 μm. The charge transport property was evaluated using the μ-τ model spectral fitting method. The average μ-τ products for electrons and holes were 2.5×10^{-3} and 5.0×10^{-4} cm^2/V, respectively. The energy resolution of the fabricated detectors from the whole wafer showed quite uniform characteristics, resulting that the average and the standard deviation were 5.3 keV and less than 6% of the average, respectively. The CdTe pixel detector was flip-chip bonded with the readout ASIC. The hybrid modules were assembled to fabricate CdTe X-ray FPD. The CdTe detector showed the very large potential as the imager of non-destructive testing and so on.

REFERENCES

1. R.Triboulet, Y.Marfaing, A.Cornet, and P.Siffert, J.Appl. Phys. **45**, No.6, 2759(1974)
2. F.V.Wald and G.Entine, Nucl.Intr.Meth., **150**,13(1978)
3. M.Ohmori, Y.Iwase, and R.Ohno, Mat. Sci.Eng., **B16**, 283(1993)
4. M.Funaki, T.Ozaki, K.Satoh and R.Ohno, Nucl. Instr. and Meth., **A436**, 120(1999)
5. H.Shiraki, M.Funaki, Y.Ando, S.Kominami, K.Amemiya and R.Ohno, 2008 IEEE Nuclear Science Symposium Conference Record, NSS'07, IEEE, 1783(2007)
6. G.Sato, T.Takahashi, M.Sugiho, M.Kuroda, T.Mitani, K.Nakazawa, Y.Odaka and S.Watanabe, IEEE trans. Nucl. Sci., **49**, 258(2002)
7. C.Matsumoto, T.Takahashi, K.Takizawa, R.Ohno, T.Ozaki and K.Mori, IEEE Trans. Nucl.Sci., **45**, 428(1998)
8. M.Tamaki, Y.Mito, Y.Shuto, T.Kiyuna, M.Yamamoto, K.Sagae, T.Kina, T.Koizumi and R.Ohno, 2008 IEEE Nuclear Science Symposium Conference Record, 2008, R09-3 (2008)

Mater. Res. Soc. Symp. Proc. Vol. 1164 © 2009 Materials Research Society   1164-L03-03

Real-time Imaging of the Electric field Distribution in CdZnTe at low temperature

P.J. Sellin[1], G. Prekas[1], A. Lohstroh[1], M. Ozsan[1], V. Perumal[1], M. Veale[1,2], P. Seller[2]

[1]Department of Physics, University of Surrey, Guildford GU2 7XH, UK
[2]STFC Rutherford Appleton Laboratory, Harwell Science and Innovation Campus, Didcot OX11 0QX, UK

ABSTRACT

Real time imaging of the electric field distribution in CZT at low temperature has been carried out using the Pockels electro-optical effect. CZT detectors have been observed to show degraded spectroscopic resolution at low temperature due to so-called 'polarization' phenomena. By mounting a CZT device in a custom optical cryostat, we have used Pockels imaging to observe the distortion of the electric field distribution in the temperature range 240K - 300K. At 240K the electric field has a severely non-uniform depth distribution, with a high field region occupying ~10% of the depth of the device under the cathode electrode and a low field in the remainder of the device. Using an alpha particle source positioned inside the vacuum chamber we have performed simultaneous alpha particle transient current (TCT) measurements. At low temperatures the alpha particle current pulses become significantly shorter, consistent with the reduced electron drift time due to a non-uniform electric field. These data provide useful insights into the mechanisms which limit the spectroscopic performance of CZT devices at reduced temperature.

INTRODUCTION

In this paper we present a study of the non-uniform electric field distribution in cadmium zinc telluride (CZT) radiation detectors at low temperatures. For many years CZT had been actively developed for use in high resolution portable gamma ray and X-ray detectors, and it has a number of desirable properties including good quantum efficiency for gamma rays and X-rays, good charge transport properties, and the ability to operate at, or close to, room temperature [1-3]. However various groups have reported stability issues in CZT, such as the buildup of a non-uniform electric field when devices are operated under irradiation with high X-ray fluence. Such 'polarization' phenomena are believed to be associated with the accumulation of trapped space charge in the detector active volume which produces an internal electric field inside the detector. This internal field tends to oppose the externally applied field. Polarization effects generally cause a deterioration of the spectroscopic performance of the detector over time, with a shift of photopeaks to lower energy and a widening of peak widths [4-6].

The spectroscopic performance of CZT is known to improve with moderate cooling of the detector (eg. to 0 °C) due to the reduction in thermal leakage current in the device. It can therefore be advantageous to mount portable CZT spectrometers onto miniature solid-state coolers such as Peltier devices. However lowering the temperature of a compensated semiconductor such as CZT can cause more complex effects due to the presence of deep states and charge traps in the material [7-8]. The position of the Fermi level in the material, and the occupation of the mid-gap states, have a dynamic and complex inter-play that depends on the rate of charge injection into the device, the capture cross-section and concentration of the various traps, and through temperature the thermal emission rate of charge out of the trapping states. These phenomena can cause the onset of polarization effects in CZT when the device temperature is reduced below approximately -40 °C., even when irradiated with low intensity radioisotope sources. This effect is similar in nature to high-flux polarization phenomena at room temperature, and may be due to space charge buildup in the material caused by reduced thermal emission of trapped charge.

In this work we use a combination of alpha particle spectroscopy to investigate the charge transport properties of the CZT devices, and electro-optical 'Pockels' imaging to directly probe the electric field distribution in our detectors as a function of device temperature. Alpha particle spectroscopy is a useful tool to investigate the charge transport properties of either electrons or holes in CZT, depending on the detector bias polarity used. For the more common case of cathode irradiation, the signals observed from the detector are due primarily to electron transport across the device. Based on the Schockley-Ramo theorem, the measured pulse height amplitude or charge collection efficiency (CCE) depends on the applied voltage V according to the Hecht equation, given by

$$CCE \approx \frac{\mu \tau_e V}{d^2}\left(1 - \exp\left(\frac{-d^2}{\mu \tau_e V}\right)\right)$$ [1]

where d is the detector thickness, and $\mu \tau_e$ is the mobility-lifetime product for electrons.

Electric field induced birefringence, commonly referred to as the Pockel's opto-electric effect, is a non-invasive method to measure the electric field distribution within bulk semiconductor crystals. The method has been applied by several groups for room temperature electric field imaging in both CdTe and CZT detector devices [9-12], and in this work we extend the technique to measure electric field distributions in CZT at low temperatures, down to 200K. In this technique the sample is illuminated with low intensity linearly polarized sub-band gap IR

16

light (980nm) such that the plane of polarization is oriented at 45^0 with respect to the applied electric field. After passing though the crystal the change in polarization of the light is detected using an analyzing polarizer at -45^0 and imaged in terms of intensity variations by a CCD camera. Under these conditions the intensity of the transmitted light $I(x,y)$ is realted to the electric field strength $E(x, y)$ by:

$$I(x, y) = I_0(x, y)\sin^2\left[\frac{\sqrt{3}\pi n_0^3 r_{41} d}{2\lambda_0} E(x, y)\right] \qquad [2]$$

where $I_0(x, y)$ is the maximum intensity transmitted with parallel polarizers and no bias on the detector, $n_0 = 2.8$ is the field-free refractive index, $r_{41} = 4.5 \times 10^{-12} mV^{-1}$ is the linear opto-electric coefficient for CZT, d is the optical path length and $\lambda = 980nm$ is the free space wavelength of the IR light.

Knowing the electric field profile as a function of distance from the electrode, for a given set of physical parameters, allows us to infer the space charge distribution in the detector $\rho(x)$ based on Poisson's equation:

$$\frac{dE}{dx} = \frac{\rho(x)}{\varepsilon} \qquad [3]$$

EXPERIMENT

Three CZT detectors were used in this study, supplied by various vendors. For each device the CZT material was of spectroscopic quality, with a room temperature electron mobility-lifetime product in the range $(1 - 9) \times 10^{-3} cm^2/Vs$. The details of the detector devices, CZT1 to CZT3, are shown in Table 1. Devices CZT1 and CZT3 were fabricated by the vendors, using standard metal contact processing. Device CZT2 was fabricated in our laboratory since it required 6-side polishing to a mirror-flat optical finish for the Pockel's optical imaging. For this device the metal contact was fabricated using thermally-evaporated gold, which extended to the edge of the device.

Alpha particle and gamma ray spectroscopy was performed on the detectors using a [241]Am radioisotope source. For alpha particle measurements the detector was mounted on an alumina ceramic substrate within a vacuum cryostat, connected to a cold finger. The temperature of the sample was controlled by liquid nitrogen flow through the cold finger, with the detector temperature variable in the range 200 - 300 K. Alpha particle pulses from the detector were processed using an Ortec 142 charge sensitive preamplifier located outside the cryostat, connected to a standard spectroscopy amplifier and multi channel analyzer. Measurement of alpha particle peak centroid was converted into charge collection efficiency (CCE), which was plotted against applied bias to produced a Hecht plot at different temperatures. For spectroscopy using 59 keV gamma rays a lower noise readout system was used which consisted of an Amptek A250 preamplifier positioned inside the vacuum chamber adjacent to the detector. The output from the preamplifier was taken outside the vacuum chamber to an external spectroscopy amplifier and multi channel analyzer. The shaping time of the spectroscopy amplifier was adjustable in the range 0.5 µs – 10 µs, and a minimum shaping time value was selected that prevented ballistic deficit.

17

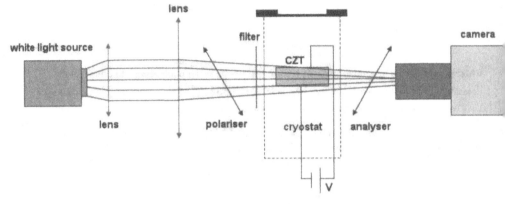

Figure 1: Schematic of the CZT Pockels imaging system

For the electro-optical field measurements the sample was mounted in an optical cryostat, also connected to a cold finger containing liquid nitrogen. The Pockels electric field measurements were carried out by imaging through the side of the crystal using plane polarized near infra-red light which had passed through an analyzing filter. The resulting image recorded the degree of rotation of polarization of the IR light in the sample, which is related to the bulk electric field strength by Eq. 1. Figure 1 shows a schematic view of the Pockels imaging system, in which a narrow bandpass (980nm) filter is placed in front of the sample to define the incident radiation and to prevent above-bandgap light being incident on the sample. The intensity of the lamp is reduced to a very low level to ensure that the electric field in the detector is not disturbed or otherwise modified by the incident light. A more complete description of the Pockels experimental setup is provided in reference [13].

Alpha particle current measurements were also carried out using the transient current technique (TCT) on device CZT2 whilst mounted in the Pockels optical chamber. In the TCT method [14] the current pulse induced in the detector from an alpha particle interaction is measured using a high bandwidth RF amplifier. In our system the amplifier used was a FEMTO current amplifier with a bandwidth of 14 MHz and a signal gain of 10^5 V/A. Each current pulse was digitized using an oscilloscope, and the width of the pulse measured to assess the drift time of the charge carriers through the device which is proportional to of the electric field strength in the detector. This measurement provides an independent cross-check of the electric field strength which can be cross compared with that obtained from the Pockels imaging.

	CZT1	CZT2	CZT3
Material supplier	eV Products	Yinnel	Redlen
Thickness	2 mm	6 mm	4.8 mm
Contact Area	2 x 2 mm	6 mm	9.8 x 4.7 mm

Table 1: Summary of CZT detector dimensions

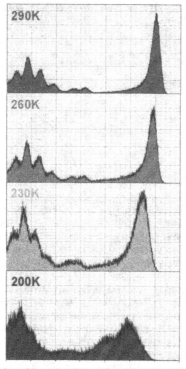

Figure 2: Gamma ray spectra from [241]Am acquired using CZT detector #1 at different temperatures. In each case the applied bias was 200V, equivalent to a mean electric field strength of 1 kV/cm, with a shaping time of 2 μs.

RESULTS and DISCUSSION

Initial spectroscopy measurements of low energy gamma rays were carried out using device CZT1 in order to investigate the performance of the detector at reduced temperature. Figure 2 shows a sequence of gamma ray spectra acquired from [241]Am using the low noise Amptek A250 preamplifier, with the detector cooled from 290K down to 200K. At 290K the spectroscopic performance of the detector is reasonable, with the 59.5 keV photopeak showing a FWHM of 3.6 keV (6.0 %). This overall energy resolution is a combination of the statistical resolution of the detector and the resolution of the preamplifier and readout system (measured using a pulser as 1.7 keV FWHM). When the detector is slightly cooled to 260K there is very little noticeable effect on the [241]Am spectrum, with the main photopeak resolution unchanged at 3.6 keV FWHM. However when cooling to 230K and 200K there is marked deterioration in the spectroscopic performance of the detector, with the resolution broadening to 5.8 keV FWHM and 9.6 keV FWHM, respectively.

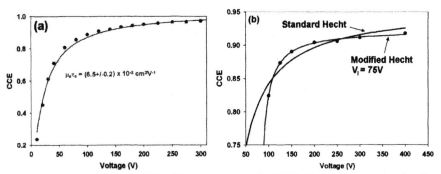

Figure 3: Alpha particle CCE vs bias voltage scans for CZT3, (a) room temperature data fitted with an unmodified Hecht equation (b) data acquired at 210K, fitted with a modified Hecht equation with a voltage offset parameter V_0

Alpha particle spectroscopy was carried out on device CZT2 to assess the electron mobility lifetime product, and any changes to the charge transport at to reduced temperature. Alpha particles from [241]Am were used to irradiate the cathode (top surface) of the device whilst in vacuum, and the resulting pulse height spectra were recorded as a function of bias voltage. The peak centroid was calibrated in terms of charge collection efficiency (CCE) using an absolute calibration from precision pulser and a known test capacitance. Figure 3(a) shows the resulting Hecht plot of alpha particle CCE as a function of voltage which was acquired at room temperature. The Hecht equation (Eq. 1) has been fitted to the data, giving a value for $\mu\tau_e$ of (6.5 ± 0.2) x 10^{-3} cm^2V^{-1}. The detector was then cooled to 210K and the same sequence of measurements were repeated, as shown in Figure 3(b). In this case the alpha particle CCE data has a form that is strongly deviating from Eq. 1. In particular there is a threshold voltage which is required before the CCE starts to increase above zero. At higher bias voltages the CCE

20

approaches a maximum value of ~92%. A modification to the standard Hecht equation is required to accurately fit this data, using a voltage offset term V_0 which represents the threshold voltage at zero CCE:

$$CCE \approx \frac{\mu\tau_e(V-V_0)}{d^2}\left(1-\exp\left(\frac{-d^2}{\mu\tau_e(V-V_0)}\right)\right)$$ [3]

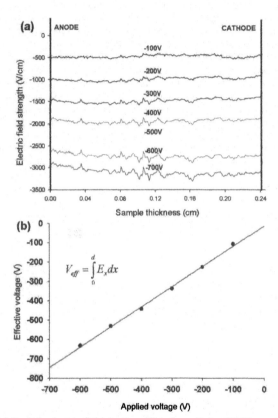

Figure 4: (a) Pockels electric field depth profiles from device CZT2 acquired at room temperature for different applied voltages, (b) the resulting linearity plot of effective voltage vs. applied voltage.

Figure 3(b) shows the fit to the data obtained using Eq. 3, obtained with an offset voltage of $V_0 = 75$V. The physical interpretation of this parameter is consistent with an internal potential in the device, with an opposite polarity to the applied bias. The external voltage applied to the detector must therefore exceed the threshold voltage V_0 before the device becomes active.

Pockels imaging was carried out on device CZT2 using low intensity plane-polarized light which had passed through the narrow-band 980 nm filter. Figure 4(a) shows a set of room temperature electric field profiles that have been extracted from the Pockels images, which demonstrate the uniform electric field distribution as a function of depth through the device. The generally flat field distributions confirm that lack of space charge accumulation in the device under normal room temperature operation. The linearity and calibration of our optical imaging system is confirmed in Figure 4(b), which shows the integral of the measured electric field profiles plotted as a function of applied bias. The data show an excellent linear response between the calculated effective voltage and the applied voltage, with a relative gain of 0.99.

Pockels imaging was performed on CZT2 whilst the device was cooled to different temperatures, down to 240K. A significant distortion to the previous uniform field distribution was observed, as shown in Figure 5. With a reduction in temperature of 20 degrees to 280K there is a significant slope in the field distribution, with the field reduced to <500 V/cm at the anode, indicating some initial build-up of space charge in the device. As the temperature is reduced further the field developed a strong bi-modal distribution with an enhanced field strength immediately under the cathode, and a low field region in the remainder of the device. At 240K the high field region is restricted to only ~10% of the depth, at the cathode, which causes a significant degradation in the detector's spectroscopic performance. Figure 6 shows the space charge concentration calculated from the integral of the measured electric field profiles, using equation 3. The data confirms the presence of a high concentration of positive space charge close to the cathode, which is consistent with increased hole trapping as intermediate-level traps freeze out at reduced temperatures.

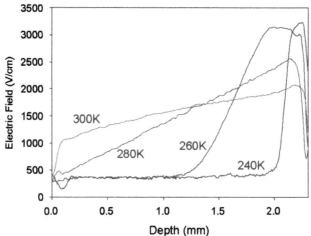

Figure 5: Pockels electric field profiles in device CZT2 acquired at temperatures between 240K and 300K. The detector was biased such that the 'left' of the profile is the anode, and the 'right' of the profile is the cathode.

22

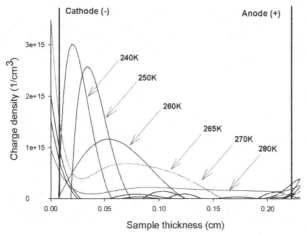

Figure 6: Space charge distributions calculated from the experimental electric field profiles in device CZT2.

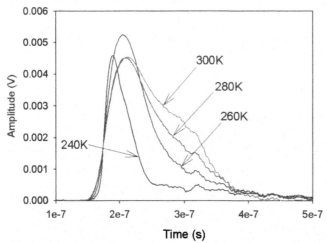

Figure 7: Alpha particle current pulses acquired from device CZT2, using the TCT method. The data show typical single pulses, acquired at temperatures between 260K and 300K.

To confirm the low temperature distortion of the internal electric field alpha particle TCT current pulse measurements were performed. The ^{241}Am alpha particle source was located inside the vacuum chamber above the device such that alpha particles were incident on the cathode. Individual current pulses from the RF amplifier were captured on a digital oscilloscope and analysed offline. The duration of each pulse gives a direct measure of the drift time of electrons through the device, which is inversely proportional to the electric field strength and the drift

23

mobility. Figure 7 shows a sequence of alpha particle induced current pulses obtained from device CZT2 as a function of temperature. With a constant applied bias the electron drift time decreases from ~160ns at 300K to ~50ns at 240K. At low temperature the shape of the current pulse matches the observed field profile - with an initial large amplitude due to high drift velocity, followed by a long low-amplitude part of the pulse corresponding to low drift velocity.

CONCLUSION

Pockels imaging is a useful tool for direct visualization of bulk field profiles in CZT detectors. By performing Pockels imaging at low temperature we have observed a strong distortion of the electric field distribution at temperatures below 280K which leads to degraded resolution for gamma ray spectroscopy applications. The Pockels data show a strongly non-uniform field distribution at 240K, with a high field region extending only ~10% under the cathode. This non-uniform field distribution has been confirmed by carrying out simultaneous alpha particle current pulse measurements that provide an independence measure of carrier drift time. The alpha particle pulse shapes show a significant reduction in drift time at low temperatures, consistent with the collapse of the high-field region in the device. The onset of the non-uniform field distribution is observed for temperatures below 280K which imposes a practical limitation on cooling of CZT devices for high resolution spectroscopy applications.

REFERENCES
[1] R. James et al., Semiconductors for Room Temperature Nuclear Detection Applications, Academic Press, New York, 1995, pp. 384.
[2] T.E Schlesinger et al., Mater. Sci. Eng. 32 (2001) 103-189.
[3] Cs. Szeles et al., IEEE Trans. Nucl. Sci. 54 (2007) 1350–1358.
[4] Derek S. Bale et al., Applied Physics Letters 92 (2008) 082101.
[5] J. Franc et al., IEEE Transactions on Nuclear Science, 54 (2007) 1416-1420.
[6] G.S Camarda et al., IEEE Nuclear Science Symposium Conference Record, Honolulu Hawaii, 2007, pp 1798-1804.
[7] B.W.Sturm et al., IEEE Transactions on Nuclear Science, 52 (2005) 2068-2075.
[8] H. Toyama et al., Japanese Society of Applied Physics, 45 (2006) 8842-8847.
[9] P. De Antonis et al., IEEE Transactions on Nuclear Science, 43 (1996) 1487-1490.
[10] H.W.Yao et al., Mater. Res. Soc. Symp. Proc. 487 (1997) 51-57.
[11] A. Cola at al., Nuclear Instruments and Methods in Physics Research A 568 (2006) 406-411.
[12] A.Burger et al., Journal of Electronic Materials, 32 (2003) 756-760.
[13] G. Prekas et al., IEEE Transactions on Nuclear Science, in press
[14] J. Fink et al., Nuclear Instruments and Methods in Physics Research A 565 (2006) 227-223.

Materials I

Mater. Res. Soc. Symp. Proc. Vol. 1164 © 2009 Materials Research Society   1164-L04-04

Differences in the Surface Charging at the (100) and (110) Surfaces of Li$_2$B$_4$O$_7$

David Wooten,[1] I. Ketsman,[2] Jie Xiao,[2] Ya. B. Losovyj,[2,3] J. Petrosky,[1] J. McClory,[1] Ya. V. Burak,[4] V.T. Adamiv,[4] and P.A. Dowben[2]

[1]Air Force Institute of Technology, 2950 Hobson Way, Wright Patterson Air Force Base, OH 45433-7765, U.S.A.

[2] Department of Physics and Astronomy and the Nebraska Center for Materials and Nanoscience, University of Nebraska-Lincoln, P.O. Box 880111, Lincoln, NE 68588-0111, U.S.A.

[3] J. Bennett Johnston Sr. Center for Advanced Microstructures and Devices, Louisiana State University, 6980 Jefferson Highway, Baton Rouge, LA 70806, U.S.A.

[4] Institute of Physical Optics, Dragomanov 23, Lviv 79005, Ukraine.

ABSTRACT

From angle resolved photoemission, the (100) surface termination of Li$_2$B$_4$O$_7$ is significantly more polar than the (110) surface termination although the accepted dipole orientation of this pyroelectric crystal is along (001). Consistent with the surface termination, the surface charging at the surface of (100) is significantly greater than observed at (110) and plays a role in the surface photovoltage effects. Because of the different interfaces formed, device properties likely depend upon crystal faces of lithium borate.

INTRODUCTION

With the fabrication of a semiconducting boron carbide, a material suitable for the fabrication of solid state neutron detectors [1-7], there has been a resurgence in the development of boron based semiconductors for neutron detection. In addition to the boron carbides [1-7], possible boron based semiconductors for solid state neutron detectors include boron nitrides [8,9], boron phosphides [10-11], and the lithium borates [12,13]. Although Li$_2$B$_4$O$_7$ lithium borate has a much larger (6.3 to 9.3 eV) band gap [14,15] than the boron carbides, boron nitrides, or boron phosphides, this class of materials has distinct advantages. The lithium borates are typically both translucent (when undoped) and can be isotopically enriched to a high degree. Li$_2$B$_4$O$_7$ can be isotopically enriched to 95 at% ^{6}Li and 97.3 at% ^{10}B [13] from the natural 7.4 at% ^{6}Li and 19 at% ^{10}B.

Since lithium tetraborate (Li$_2$B$_4$O$_7$) single crystals are pyroelectric and piezoelectric [16], surface termination and interface effects must also be seriously considered in solid state device design. This is a critical consideration as doping will serve to enhance differences along different crystal directions [17]. Doping of the Li$_2$B$_4$O$_7$ is certainly possible [13], and essential for solid state neutron detection applications, given the very high resistivities of undoped crystals. These undoped resistivities are on the order of 10^{10} $\Omega\cdot$cm [12].

EXPERIMENT

The $Li_2B_4O_7$ single crystals were grown from the melt by the Czochralski technique as described elsewhere [13,15]. Clean surfaces were prepared by several methods including resistive heating and combinations of sputtering and subsequent annealing. The electronic structure and stoichiometry were similar in all cases.

Combined ultraviolet photoemission (UPS) and inverse photoemission (IPES) spectra were used to characterize the placement of both occupied and unoccupied states of $Li_2B_4O_7(100)$ and $Li_2B_4O_7(110)$ single crystals. The temperature-dependent angle-resolved photoemission experiments were performed using the 3 m toroidal grating monochromator (3 m TGM) beam line [18] in an ultra high vacuum (UHV) chamber previously described [18-20]. Photoemission was conducted over a range of temperatures from 250 to 700 K, but the data presented here were taken at 623 K, where surface charging was found to be negligible, unless noted otherwise. While the light incidence angle was varied, to change the orientation of the $\underline{E}$ of the plane-polarized incident light, the photo-electrons were collected along the surface normal throughout this work to preserve the high point group symmetry ($\bar{\Gamma}$). The inverse photoemission (IPES) spectra were obtained by using variable energy electrons incident along the sample surface normal, again to preserve the high point group symmetry ($\bar{\Gamma}$), while measuring the emitted photons at a fixed energy (9.7 eV) using a Geiger-Müller detector with an instrumental linewidth of about 400 meV [21-22]. The inverse photoemission spectra were taken for a sample temperature of 300 K. Checks to the placement of the Fermi level in both the angle-resolved photoemission and inverse photoemission experiments were performed using tantalum films in electrical contact with the samples. Surface charging shifts in the photoemission were accounted for using the Li 1s and O 2s shallow core levels.

DISCUSSION

The $Li_2B_4O_7(100)$ and $Li_2B_4O_7(110)$ valence and conduction band density of states:
From Figure 1, it is clear that the band gaps for the $Li_2B_4O_7(100)$ and the $Li_2B_4O_7(110)$ single crystal surfaces are about 9 to 9.5 eV from the combined photoemission and inverse photoemission for the high symmetry $\bar{\Gamma}$ point of each surface.

28

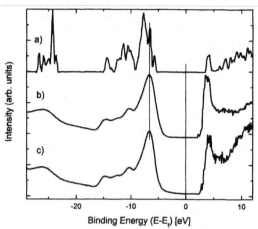

Binding Energy (E-E$_r$) [eV]

Figure 1. The combined experimental photoemission and inverse photoemission results for solid Li$_2$B$_4$O$_7$ with theory: (a) the theoretical density of states of solid Li$_2$B$_4$O$_7$ abstracted from the work of Islam et al. [14]; (b) combined experimental photoemission and inverse photoemission results for Li$_2$B$_4$O$_7$(110) surface, taken with the in-plane $\underline{E}$ vector oriented along [1$\underline{1}$0], and (c) Li$_2$B$_4$O$_7$(100) surface, taken with the in-plane $\underline{E}$ vector oriented along [011]. The photoemission was taken at a photon energy of 56 eV with electrons collected along the surface normal, while the inverse photoemission was taken with electrons incident normal to the sample.

This tends to be towards the higher end of the theoretically predicted band gaps that ranging from 6.2 to 9.7 eV [14-15]. This is surprisingly good agreement given that local density approximation (LDA) models will underestimate the band gap [23]. The experimental estimate of the band gap could also be flawed for several reasons as well: the data is shown for only a limited wave vector sample (not averaged for the entire bulk Brillouin zone), charging effects that remain (not removed in the binding energy corrections to the data, in Fermi level placement calibrations) will tend to increase the apparent band gap. Furthermore photoemission and inverse photoemission are final state [24], not initial state spectroscopies, and very surface sensitive.

After corrections for photovoltaic charging (as discussed below), we place the shallow core level binding energies at 56.7±0.5 eV for the Li 1s and 26.0±0.6 eV for the oxygen 2s. These tend to be somewhat higher values than reported previously for lithium tetraborate Li$_2$B$_4$O$_7$ [14,25]. Without adequate preparation of the stoichiometric clean surface, we have reason to believe that surface contributions will dominate even the core level photoemission spectra and lead to an artificial decrease in the core level binding energies.

In spite of the numerous deficiencies of the combined photoemission and inverse photoemission experiments, both Li$_2$B$_4$O$_7$(100) and Li$_2$B$_4$O$_7$(110) exhibit a density of states that qualitatively resembles that expected from the model bulk band structure of Li$_2$B$_4$O$_7$ [14,15], as seen in Figure 1. Surface contributions, nonetheless, cannot be neglected.

The effect of Li₂B₄O₇(100) and Li₂B₄O₇(110) surface termination:

The effect of $Li_2B_4O_7(100)$ and $Li_2B_4O_7(110)$ surface termination:

From Figure 2, it is clear that the light polarization dependent photoemission differs for $Li_2B_4O_7(100)$ and $Li_2B_4O_7(110)$.

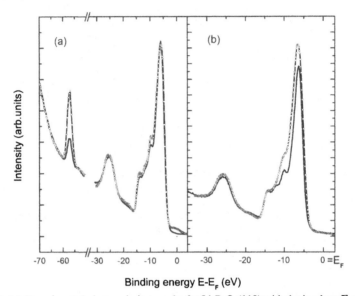

Figure 2. (a) Experimental photoemission results for $Li_2B_4O_7(110)$ with the in-plane $\underline{E}$ of the incident light oriented along [001] and for (b) $Li_2B_4O_7(100)$ with the in-plane $\underline{E}$ of the incident light oriented along [010]. The photon energy is 95 eV; the orientation of light incidence $\underline{E}$ is 45 degrees {-o-o-o-} or 70 degrees {——} with respect to surface normal.

The valence band density of states, as indicated by the angle-resolved photoemission, is strongly enhanced for $Li_2B_4O_7(100)$, with incident light polarization placing the electric vector $\underline{E}$ more along the surface normal (Figure 2b). This is not the case for $Li_2B_4O_7(110)$, where there is little influence of incident light polarization on the angle resolved photoemission (Figure 2a), except for the Li 1s shallow core as is generally expected for an Li 1s shallow core with a strong surface contribution.

The enhancement of the photoemission spectra for $Li_2B_4O_7(100)$, with the electric vector $\underline{E}$ of the incident light polarization placed more along the surface normal, is indicative of a polar termination for this surface. The surface dipoles of the A_1 irreducible representation are placed along the surface normal, while for $Li_2B_4O_7(110)$, these surface dipoles must lie more in the plane of the surface. From the comparison with theory [14], the photoemission must be largely dominated in the valence band region by boron and oxygen 2p weighted bands. So these surface dipoles for $Li_2B_4O_7(100)$ are related to the oxygen and boron p_z weighted bands [14].

30

Surface Photovoltaic Charging:

Figure 3 shows that the surface photovoltage charging, as observed in photoemission, also differs for $Li_2B_4O_7(100)$ and $Li_2B_4O_7(110)$.

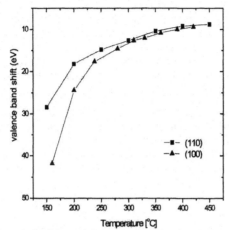

Figure 3. Position of the valence band maximum for both the (110) and (100) surfaces of $Li_2B_4O_7$, as a function of temperature. The data was obtained from photoemission spectra taken at a photon energy of 95 eV, with the photoelectrons collected normal to the surface. There is hysteresis, so data is shown only for increasing, not decreasing temperatures.

The shift in the valence band maximum with temperature is much greater for $Li_2B_4O_7(100)$ than for $Li_2B_4O_7(110)$. In many such cases, the surface voltaic charging can be dominated by surface conductivity. The greater charging observed for the $Li_2B_4O_7(100)$ surface over the $Li_2B_4O_7(110)$ surface is consistent with the differences in surface termination. As expected, the more polar surface is the better dielectric than the more non-polar surface.

CONCLUSIONS

We see that although the bulk dipole direction for $Li_2B_4O_7$ is generally accepted to be along the [001] direction [16], the other surface terminations can result in distinct polar surface terminations. These complications of surface termination become a concern for making devices, as the different surface terminations can and will mean different interface states depending on the contact electrodes used to fabricate the $Li_2B_4O_7$ device.

ACKNOWLEDGMENTS

The authors acknowledge insightful discussions with K. Belashchenko. This work was supported by the Defense Threat Reduction Agency (Grant No. HDTRA1-07-1-0008), and the Nebraska Research Initiative. This work was undertaken in partial fulfillment of the degree at AFIT by one author (DW). The views expressed in this article are those of the authors and do not reflect the official policy or position of the Air Force, Department of Defense or the U.S. Government.

REFERENCES

1. A.N. Caruso, Ravi B. Billa, S. Balaz, Jennifer I. Brand, and P.A. Dowben, *J. Phys. Cond. Mat.* **16**, L139 (2004).

2. B.W. Robertson, S. Adenwalla, A. Harken, P. Welsch, J.I. Brand, P.A. Dowben, and J.P. Claassen, *Appl. Phys. Lett.* **80**, 3644 (2002).

3. B.W. Robertson, S. Adenwalla, A. Harken, P. Welsch, J.I. Brand, J.P. Claassen, N.M. Boag, and P.A. Dowben, *Adv. Neutron Scatt. Instrum.*, Anderson, I.S.; Guérard, B., Eds. *Proc. SPIE* **4785**, 226 (2002).

4. S. Adenwalla, R. Billa, J.I. Brand, E. Day, M.J. Diaz, A. Harken, A.S. McMullen-Gunn, R. Padmanabhan, and B.W. Robertson, *Penetrating Radiation Systems and Applications V, Proc. SPIE* **5199**, 70 (2003).

5. K. Osberg, N. Schemm, S. Balkir, J.I. Brand, S. Hallbeck, P. Dowben, and M.W. Hoffman, *IEEE Sensors J.* **6** (2006) 1531; K. Osberg, N. Schemm, S. Balkir, J.I. Brand, S. Hallbeck, P. Dowben, *2006 IEEE International Symposium on Circuits and Systems (ISCAS 2006) Proceedings* 1179 (2006).

6. A.N. Caruso, P.A. Dowben, S. Balkir, N. Schemm, K. Osberg, R.W. Fairchild, O. B. Flores, S. Balaz, A.D. Harken, B.W. Robertson, and J.I. Brand, *Mat. Sci. Engin.* B **135**, 129 (2006).

7. E. Day, M.J. Diaz, and S. Adenwalla, *J. Phys. D: Appl. Phys.* **39**, (2006) 2920

8. D.S. McGregor, T.C. Unruh, W.J. McNeil, *Nucl. Instrum. Methods Phys. Res.* A **591**, 530 (2008)

9. J. Uher, S. Pospisil, V. Linhart, M. Schieber, *Appl. Phys. Lett.* **90**, 124101 (2007)

10. Y. Kumashiro, *J. Mater. Res.* **5**, 2933 (1990)

11. Y. Kumashiro, *J. Solid State Chem.* **133**, 314 (1997)

12. Sangeeta, K. Chennakesavulu, D.G. Deai, S.C. Sabharwal, M. Alex, M.D. Ghodgaonkar, *Nucl. Instrum. Methods Phys. Res.* A **571**, 699 (2007)

13. Ya. B. Burak, V.T. Adamiv, I.M. Teslyuk, V.M. Shevel, *Rad. Meas.* **38**, 681 (2004)

14. M.M. Islam, V.V. Maslyuk, T. Bredow, C. Minot, *J. Phys. Chem.* B **109**, 13597 (2005)

15. V.T. Adamiv, Ya. B. Burak, I.V. Kityk, J. Kasperczyk, R. Smok, M. Czerwinski, *Optical Materials* **8**, 207 (1997)

16. A.S. Bhalla, L.E. Cross, R.W. Whatmore, *Jap. J. Appl. Phys.* **24**, suppl. 24-2, 727 (1985)

17. P.A. Dowben A. Miller, editors, Surface Segregation Phenomena, *CRC Press*, Boca Raton, Florida (1990)

18. Ya.B. Losovyj, I. Ketsman, E. Morikawa, Z. Wang, J. Tang, P.A. Dowben, *Nucl. Instrum. Methods Phys. Res.* A **582**, 264 (2007)

19. Ya. B. Losovyj, D. Wooten, J. C. Santana, J. M. An, K. D. Belashchenko, N. Lozova, J. Petrosky, A. Sokolov, J. Tang, W. Wang, N. Arulsamy, P.A. Dowben, *J. Phys. Cond. Matter* **21** (2009) 045602

20. P.A. Dowben, D. LaGraffe, and M. Onellion, *J. Phys. Cond. Matt.* **1**, 6571 (1989)

21. D.N. McIlroy, J. Zhang, P.A. Dowben, D. Heskett, *Mat. Sci. Eng.* A **217/218**, 64 (1996)

22. C.N. Borca, T. Komesu, P.A. Dowben, *J. Electron Spectrosc. Rel. Phenom.* **122**, 259 (2002)

23. I.N. Yakovkin, P.A. Dowben, *Surf. Rev. Letters* **14**, 481 (2007)

24. J.E. Ortega, F.J. Himpsel, D. Li, P.A. Dowben, *Solid State Commun.* **91**, 807 (1994)

25. A. Yu Kuznetsov, A.V. Kruzhalov, I.N. Ogorodnikov, A.B. Sobolev, L.I. Isaenko, *Phys. Sold. State* **41**, 48 (1999)

CdTe and CdZnTe Detectors II

Mater. Res. Soc. Symp. Proc. Vol. 1164 © 2009 Materials Research Society   1164-L05-01

Development of X-ray, Gamma Ray Spectroscopic Detector Using Epitaxially Grown Single Crystal Thick CdTe Films

M. Niraula, K. Yasuda, H. Ichihashi, Y. Kai, A. Watanabe, W. Yamada, H. Oka, T. Yoneyama, K. Matsumoto, T. Nakanishi, D. Katoh, H. Nakashima, and Y. Agata
Nagoya Institute of Technology, Graduate School of Engineering
Gokiso, Showa, Nagoya 466-8555, Japan

ABSTRACT

In this paper, we review our efforts in the spectroscopic detector development using epitaxially grown thick single crystal films of CdTe. The films were grown on GaAs and Si substrates using metalorganic vapor phase epitaxy growth technique. High crystalline quality thick single crystal CdTe films (>260 μm) were obtained where the growth rates could be varied from 10 to 70 μm/h by adjusting the precursor's flow rates, ratios and the substrate temperatures. Spectroscopic detectors were fabricated in a p-CdTe/n-CdTe/n$^+$-GaAs or p-CdTe/n-CdTe/n$^+$-Si heterojunction diode structure. Both types of detector were capable of detecting and resolving energy peaks from a gamma ray source. However, the spectroscopic performance of p-CdTe/n-CdTe/n$^+$-Si detectors was better than that of the p-CdTe/n-CdTe/n$^+$-GaAs detectors. Details on the growth characteristics, detector fabrication and the detector performance are reported. Furthermore, current challenges in this detector fabrication technique are discussed.

INTRODUCTION

CdTe and CdZnTe are the most widely explored compound semiconductors for x-ray, gamma ray detector applications. Their unique physical properties such as high average atomic number and density give high detection efficiency for the impinging photons, while the wide bandgap values make them suitable for room temperature operation. Due to intensive research carried out in recent years, material properties, especially the charge transport properties of electron, have significantly improved [1-4]. Small spectroscopic detectors and small-area imaging array with high-energy resolutions are already developed and some of them are now commercially available.

Today, CdTe, CdZnTe detector crystals are grown by melt-growth techniques, such as the high-pressure or vertical Bridgman or the traveling heater method [1-5]. High-resistivity crystal required for detector fabrication is obtained by compensating background impurities and native defects with external dopants [1-3]. Unfortunately, melt-grown bulk crystals presently contain numerous crystal defects, such as grain boundaries, twins, Te-inclusions, cracks and pipes, and in some cases exhibit polycrystallinity that lead to poor electrical transport properties and inhomogeneity, which limit detector size to small dimensions. The detector grade crystal wafers are often obtained by screening and separating portions of single crystalline defect-free crystals from the large volume crystal ingots. This obviously decreases the crystal yields, thus increasing the overall detector cost, and also limits the size of the available wafers. This hampers the development of large-area and high-sensitivity, high-energy resolution imaging detectors significantly [6], [7].

The vapor-phase growth of detector grade thick single crystal films of CdTe or CdZnTe on large area substrates such as GaAs or Si offers several advantages over melt-crystal growth.

As the growth is carried out at lower substrate temperatures compared to the melt-growth, higher quality crystals with a reduced density of defects can be produced. In addition, the low temperature growth offers stoichiometric crystals as well as high-accuracy and easy control of the compensating dopants. As a result, homogeneous large-area crystals with good electrical properties can be obtained. However, growth of detector-grade single crystal epitaxial films with sufficient thickness as required for efficient absorption of the incident x-ray or gamma ray presents a considerable challenge. For a medical imaging detector working in low x-ray, gamma ray energies (from a few keV up to 100 keV), more than 500 μm thick single crystal CdTe film would be required for achieving a reasonable detection efficiency. In addition to the challenges of achieving the required thickness, there is another enormous difficulty associated with this heteroepitaxial growth. The large difference in the lattice constants between the film and the substrates (14% in CdTe-GaAs, 20% in CdTe-Si), the difference in their thermal expansion coefficients, and the difference in the nature of their chemical bonding characteristics make the growth very difficult.

We are investigating growth of high-quality, very thick single crystal films of CdTe on GaAs as well as on Si substrates using the metalorganic vapor phase epitaxy (MOVPE) technique, and development of x-ray, gamma ray detectors utilizing these films [8-11]. Dedicated epitaxial growth techniques were developed to grow high crystalline quality thick single crystal CdTe films directly on GaAs and Si substrates [10-12]. Radiation detectors were fabricated in a heterojunction diode structure comprising either p-CdTe/n-CdTe/n+-GaAs or p-CdTe/n-CdTe/ n+-Si. Both types of detectors exhibited good diode-like characteristics and were capable of detecting and resolving energy peaks from a gamma ray source ([241]Am) at room temperature. We will present the detailed growth characteristics on the different substrates, detector fabrication and then discuss how the detector structure influences the detector's charge transport properties. Furthermore, current challenges in this detector development technique will be discussed.

EXPERIMENT

Epitaxial growth on GaAs substrates

The growth was carried out at atmospheric pressure in a vertical-type MOVPE reactor. The reactor was custom-designed, which has been described in detail elsewhere [12]. Dimethylcadmium (DMCd) and diethyltellurium (DETe) were used as the precursor. The reactor wall was set at a temperature of 200 °C, and the substrate temperature was varied from 415 to 560 °C, during the growth. (100) semi-insulating (SI) GaAs substrates were used in this investigation. In order to obtain (100) CdTe growth consistently on (100) GaAs, the GaAs substrates, after the usual chemical etching, were annealed in a hydrogen atmosphere prior to the growth, and Te-precursor was introduced first while initiating the growth. The details on the initial growth mechanism of CdTe on GaAs substrates can be found in [13,14]. Film orientation and the crystallinity were examined by the x-ray diffraction (XRD) measurement using Cu K_α as x-ray source. X-ray double-crystal rocking curve (DCRC), and low temperature (4.2 K) photoluminescence (PL) measurement using excitation from an Ar^+-ion laser (514.5 nm) were further used to evaluate the crystalline quality of the films.

Epitaxial growth on Si substrates

Epitaxial growth on (211) Si substrate was carried out in a vertical-type MOVPE reactor capable of accommodating 4-in Si wafers. The growth was carried out at a substrate temperature of 560 or 600 °C, using DMCd and DETe precursors. Challenges associated with the direct growth of high-quality single crystal CdTe films on Si substrate were overcome by employing a special Si substrate pre-treatment method. This was achieved by annealing (211) Si substrates at 800-900 °C with pieces of GaAs in a hydrogen environment in a separate horizontal chamber as described in [10,11]. This annealing process removes the native oxide from the Si substrate and deposits a thin interfacial layer of GaAs on the substrates, forming a coherent interface. The annealed substrates were then transported immediately to the growth chamber for the CdTe growth. XRD and PL were used to evaluate the crystalline quality of the films as described above.

Detector fabrication

The detectors were fabricated in a heterojunction diode structure comprising either p-CdTe/n-CdTe/n$^+$-GaAs, or p-CdTe/n-CdTe/n$^+$Si. The n-CdTe buffer layer in both types of detectors was typically 5 μm thick, and doped with iodine to an electron concentration of 10^{16}-10^{17} cm^{-3}. This layer was grown at 350 °C using ethyliodide as a dopant. Details on this detector structure and the importance of the n-CdTe buffer layer in detector charge transport property and detector performance are described elsewhere [15, 16]. On the top of n-CdTe buffer layer, 90 to 260 μm thick CdTe films were grown at substrate temperature of 560 or 600 °C. Finally, gold was evaporated on both sides of the detector as ohmic contact. The detectors were then diced into 1mmx2mm pieces, mounted inside a standard chip carrier and wire bonded. The detectors were evaluated by the current-voltage (I-V) measurement and the nuclear radiation detection tests using [241]Am gamma source.

RESULTS AND DISCUSSION

Growth characteristics on GaAs substrates

The growth of CdTe films on the SI-GaAs was investigated in the substrate temperature range from 415 to 560 °C, where the DMCd flow-rate was 7×10^{-5} mol/min and the VI/II precursor ratio was one. The growth rate increased with the substrate temperature, as shown in Fig. 1, indicating the growth process is controlled by surface kinetics in the temperature range investigated here. A growth rate of 12 μm/h was attained at the substrate temperature of 560 °C.

All films were cubic, their growth orientation was controlled to be in the (100) plane parallel to the substrates [13]. Fig. 2 shows the dependence of DCRC FWHM values of the CdTe (400) reflection plane with the thickness for the films grown at different substrate temperatures. The FWHM value rapidly decreases with increasing thickness and then remains between 50 to 70 arcsec for the films thicker than 30 μm. This indicates that highly crystalline thick CdTe films can be obtained by the MOVPE growth technique with a high growth rate.

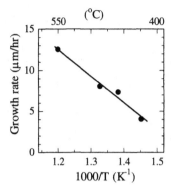

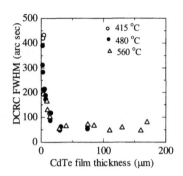

Fig. 1. Growth rate of CdTe films as a function of substrate temperatures.

Fig. 2. X-ray DCRC FWHM values of CdTe (400) plane as function of film thickness grown at different subs. temperatures.

Result from the PL measurement of the films grown at two different substrate temperatures is shown in Fig. 3. The result showed high-intensity excitonic emissions, and very small defect-related peaks (both shallow and deep DAP) from both films. The (D^0,X) emission was dominant on the 50 μm thick film grown at 415 °C, while the (A^0,X) emission was dominant for the 110 μm thick film grown at 560 °C. The (D^0,X) line is considered due to the recombination of an exciton bound to a neutral donor, probably Ga impurity diffused from the substrate in this case. The (A^0,X) emission is attributed to the recombination of an exciton bound to a neutral acceptor originating either from V_{cd} or from a complex including V_{cd} and Ga. Secondary ion mass spectroscopy (SIMS) confirmed the Ga diffusion in the films [8]. The concentration of Ga was lower in the film grown at 415 °C than that of the film grown at 560 °C, however, it was uniform throughout the depth investigated [8]. Moreover, the composition mapping showed islands of Ga on the

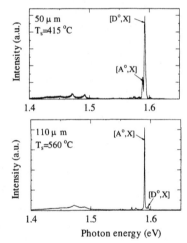

Fig. 3. 4.2 K PL spectra of CdTe films grown at 415 °C (top) and 560 °C (bottom).

surface, most probably out-diffused from the substrate through the pipe diffusion. In this regard, the difference in the emission peaks is considered due to the difference of defect complex formation by the diffused Ga. The high growth temperature (560 °C) enhances the formation of compensating complex of Ga with the Cd-vacancy, while the low-temperature growth assists the activation of the diffused Ga.

38

Resistivity measurement carried out by means of the van der Pauw method at room temperature on films grown on the SI-GaAs substrates showed that films grown at 560 °C had a p-type resistivity in the order of 10^5–10^6 Ω cm, while those grown at 415 °C had an n-type resistivity in 10^3 Ωcm order. The resistivity of the films essentially remained constant as the thickness increased above 50 μm. The resistivity of the substrate was in the order of 10^6Ω cm, which is not sufficiently large, which might have affected the measured value of the film resistivity, particularly for the films grown at higher substrate temperature.

p-CdTe/n-CdTe/n⁺-GaAs diode detector

Cross-sectional diagram of the diode detector is shown in Fig. 4 and the diode detector's *I-V* characteristics is shown in Fig. 5. The detector exhibits good rectification property, where a large current flows when a positive bias is applied to the electrode on the p-CdTe side (forward bias), but the current remains low in the opposite case (reverse bias). The dark current of the detector was 3.1 μA/cm² for an applied reverse bias of 40 V. The reverse dark current has a square-root dependence with applied bias at low bias region (up to about 10 V), but the dependence then turns to be linear. This indicates generation current from the depletion region is prominent at the low voltage, whereas the surface leakage current becomes dominant at the higher voltage range.

The radiation detection test was performed using a standard set up consisting of a pre-amplifier, a shaping amplifier and a multi-channel analyzer. The whole detector assembly was covered with a thin aluminum foil during the test. A positive bias was applied to the electrode on the n⁺-GaAs side of the detector, while the electrode on the CdTe side was grounded, and an uncollimated source of ^{241}Am radioisotope was used to irradiate the detector through this side.

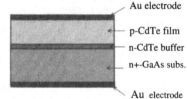

Fig. 4. Cross-sectional diagram of p-CdTe/n-CdTe/n+-GaAs diode detector.

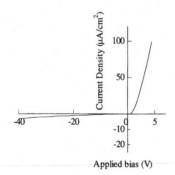

Fig. 5. Typical current-voltage characteristics of the p-CdTe/n-CdTe/n+-GaAs heterojunction diode detector at room-temperature.

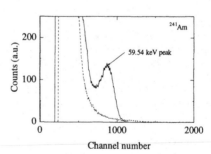

Fig. 6. Pulse height spectrum taken from a ^{241}Am gamma source at room temperature. The dotted line represents response when no gamma source was irradiated to the detector.

39

Fig. 6 shows the pulse height spectrum taken from a ^{241}Am gamma source at room temperature. The spectrum was collected by applying a bias voltage of 50V, and a shaping time of 0.5μs. Although the 59.5 keV gamma-peak is clearly resolved, other low energy peaks are not resolved because of the high level of dark current noise. The energy peak obtained is board, with a distinct low energy tail. The broad peak indicates fluctuations in the charge collection in the detector as well as the effect of the electronic noise of the detector, while tailing indicates charge loss in the detector due to the trapping or recombination. Result of the capacitance-voltage measurement showed that diode was not fully depleted in the operating voltage [16]. The undepleted region most likely causes the incomplete charge collection in the detector. Additionally, electrically active deep levels in the crystal cause poor charge transport of the carriers leading to the trapping and detrapping. Time-of-flight measurement performed in the detector confirmed the carrier (electron) travel with multiple trapping and detrapping, which restrict their charge transport. [8, 16]. At this point, it is difficult to identify, but we consider native defects (dislocations) in the crystal along with impurities present are the main factors that affect charier charge transport properties. CdTe films on GaAs substrates contain a large amount of Ga as confirmed by the SIMS [8] and may be regarded as a source of the electrically active defects. In fact, Ga-donor level and Ga-related complex (A-center) acceptor levels are reported, however, their energies are not deep enough to consider them as a deep trap [2, 17, 18]. There might be some unidentified Ga-related complex, otherwise clustering of Ga atoms around the dislocations, through pipe-diffusion, and giving rise to modified energy levels in the bandgap can be considered as another possibility that degrade the device property by enhancing the carrier recombination. The C-V data revealed that the net acceptor concentration (N_a-N_d) of the p-CdTe layer was around 10^{14}cm^{-3} [16], mainly due to the high concentration of Cd-vacancies and the Ga A-center acceptor. This, in turn, decreases the resistivity of the material and prevents the full extension of depletion layer in p-CdTe side. Achieving high resistivity p-CdTe film, and minimization of the dark current are the main concern at present in these detectors. Hence, reduction of the native defects as well as electrically active impurities in the CdTe film is an important issue in these detectors.

Growth characteristics on Si substrates and detector development

Si substrates offer advantages over GaAs as they are available in large-area, are robust and cheap. Also there will be no Ga-contamination problem as in case of the GaAs substrates. Films with different thicknesses varying from 40 to 260 μm were grown with a growth rate varying from 40 to 70 μm/h [19]. All grown films had good substrate adhesion and good surface morphology. Fig. 7 shows the XRD patterns of thick CdTe films grown on (211) Si at two different temperatures and with different growth-rates. The plot indicates that the CdTe film is monocrystalline with the growth orientation similar to that of the Si substrate. It is noteworthy that no notable peaks other than that of the CdTe (422)

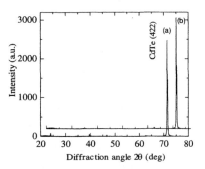

Fig. 7. XRD patterns of CdTe films on (211) Si substrates (a) 260 μm-thick, grown at 600 °C, growth rate 65μm/h, (b)100 μm-thick grown, at 560°C, growth rate 50μm/h. .

are observed, indicating no twinned regions are present in the grown film. The x-ray DCRC measurement revealed that the FWHM values from the CdTe (422) plane were in the range from 400 to 600 arcsec. There were no significant differences in the FWHM values for the films grown at different temperatures and with the different growth rates.

Fig. 8 shows a PL spectrum from 100 μm-thick film grown at a substrate temperature of 600 °C. The film was grown with a growth rate of 65 μm/h. The PL spectrum shows similar feature as that of films grown on GaAs substrates. It has sharp and strong excitonic emission line, but a very broad and weak deep level emission in 1.4 to 1.5 eV range. Sharp peak at 1.5902 eV, which is usually observed in p-type CdTe samples can be assigned as the recombination of exciton bound to neutral acceptor (A°, X), with cadmium vacancy (V_{Cd}) or its complex probably acting as neutral acceptors [20, 21]. The small but distinct peak at 1.5937 eV (Fig 8 inset) is the donor bound exciton emission (D°, X), with residual iodine in the growth chamber probably acting as a neutral donor. Weak emissions visible at 1.5958 and 1.5973 eV are due to the free exciton emissions.

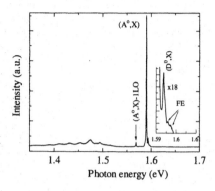

Fig. 8. 4.2K PL spectrum of a 100 μm-thick CdTe film on (211)Si. The inset shows the edge-emission band in a magnified scale.

Strong and sharp bound excitonic emission line, with a FWHM value of 0.9 meV, together with free exciton emission and weak DAP emission band indicate that grown film is of good crystalline quality.

These films exhibit a p-type conductivity, however, their exact resistivity value could not be determined by the Van der Pauw Hall measurement as the resistivity of the Si substrate was many orders lower compared to the grown film. We estimated their equivalent resistivity during current-voltage measurement of the diode as explained below. Also, it should be noted that no measurable amount of Ga or As was detected during the SIMS measurement from the CdTe films grown on Si substrates, although the Si substrates were undergone heat treatments using GaAs (see experimental section) before the growth.

Heterojunction diode-type detector was fabricated using thick epitaxial films grown on n^+-Si substrates. The cross-sectional diagram of the detector was similar to that shown in Fig. 4. In this detector fabrication also low temperature grown n-CdTe buffer layer was used to improve the junction property [22]. The p-CdTe film thickness was about 160 μm The *I-V* characteristics of the p-CdTe/n-CdTe/n^+-Si heterojunction diode detector was similar to that of

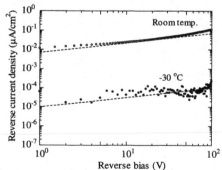

Fig. 9. Reverse current of the p-CdTe/n-CdTe/n^+-Si diode at room-temperature(24 °C) and -30 °C.

p-CdTe/n-CdTe/n$^+$-GaAs (see Fig. 5). The diode exhibits good rectification property as usual, but offers much lower dark current than that of the p-CdTe/n-CdTe/n$^+$-GaAs detector. The dark current of this detector was typically 0.11 μA/cm^2 at room temperature for an applied reverse bias of 100V, as shown in Fig. 9, which is more than an order of magnitude lower than that of p-CdTe/n-CdTe/n$^+$-GaAs detector. By cooling the detector to -30 °C, the detector dark current could be reduced by three orders of magnitude from its room-temperature value. The decrease of the dark current by cooling attributed to the decrease of diode generation current [23].

The resistivity of the p-CdTe films could not be determined by the the Van der Pauw Hall measurement. Hence, we made an estimation of their equivalent resistivity using the *I-V* data in forward bias and reverse bias direction. Using the room-temperature forward *I-V* data, the bulk resistivity was calculated approximately equal to 1.05x10^6 Ω·cm. For this calculation the diode thickness was taken as 465 μm, sum total of Si substrate thickness (300 μm) and the total CdTe film thickness (165 μm). This resistivity value only provides a rough approximation for the p-CdTe film resistivity. Again, we used the room-temperature reverse *I-V* data and calculated equivalent resistivity of the diode in the reverse direction at 70 V, which was 1.9x10^{10} Ω·cm.

Fig. 10 shows detector's spectroscopic response. The detector was fabricated by using a 110 μm thick p-CdTe film grown at 600 °C. The detector was capable of resolving gamma peaks (one main peak at 60 keV region and another less distinct peak at low energy region) at room-temperature. Though the peak broadening and low energy tailing is still present in the spectrum, however, there is a distinct improvement when compared with the spectrum recorded with the detectors based on GaAs substrates. In order to reduce the thermal noise, we cooled down the detector to -30 °C. The spectrum (Fig. 10(b)) was collected by applying a reverse bias of 300 V and a shaping time of 1 μs. The spectrum shows the detector is capable of resolving all peaks associated with the [241]Am gamma source. The energy resolution at 59.5 keV peak was about 6% at FWHM value. Several detectors fabricated from the different portions of the wafer grown with similar conditions exhibited similar detection properties.

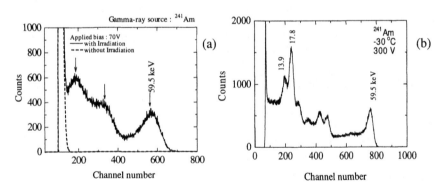

Fig. 10. Pulse height spectra taken from a [241]Am gamma source with the p-CdTe/n-CdTe/n+-Si diode detector at (a) room temperature, detector bias 70V, (b) -30 °C, detector bias 300V.

A significant improvement in performance was achieved fabricating detectors using CdTe films grown on Si substrates (p-CdTe/n-CdTe/n$^+$-Si detector) over detectors based on GaAs substrates (p-CdTe/n-CdTe/n$^+$-GaAs detector). Despite the improvements, the detectors still suffer from poor charge transport properties, specially during room temperature operation leading to poor energy resolution and low-energy tailing. Furthermore, the detection efficiency is low, and the pulse height spectrum exhibited a high background continuum due to inadequate thickness of the p-CdTe film. Another challenge associated with this heteroepitaxial system is the large number of misfit dislocations located at the n$^+$-Si and the n-CdTe junction. These misfit dislocations causes a large leakage current when the depletion layer spread towards this junction. In order to address these challenges, we are currently investigating following
- thickness and carrier concentration increment of the n-CdTe buffer layer
- reduction of net carrier concentration (N_a-N_d) of p-CdTe
- optimization of growth conditions to minimize the dislocations and crystallinity improvement
- growth of thicker p-CdTe films to increase the detection efficiency

This will help us to achieve fully depleted detectors with a low leakage current, which in turn improve the detector charge transport properties and the overall detector performance.

CONCLUSIONS

The development of spectroscopic detectors based on epitaxially grown thick single crystal CdTe films has been discussed. Using our MOVPE growth technique, more than 260 μm-thick single crystal CdTe films of high crystalline quality were obtained with a high growth rate varying from 10 to 70 μm/h. We demonstrated that these thick film based detector can be used for spectroscopy and nuclear imaging by fabricating detectors in p-CdTe/n-CdTe/n$^+$-GaAs or p-CdTe/n-CdTe/n$^+$-Si heterojunction diode structures, and detecting and resolving energy peaks from a gamma source.   This detector fabrication technique has certain advantages over current bulk-crystal based detectors as it offers large-area and uniform crystals with controlled electrical properties that are required for developing imaging arrays. However, there are certain challenges, such as crystallinity imperfection, inadequate film thickness which need to be addressed in order to improve the detector performance further.

ACKNOWLEDGMENTS

This work was supported in parts by the New Energy and Industrial Technology Development Organization (NEDO)-Japan under Grant 03A47020, by the Grant-in-Aid for Scientific Research of the Japan Society for the Promotion of Science (JSPS) (A-19200044 and B-20310095), and by the Research Foundation for the Electrotechnology of Chubu (Japan).

REFERENCES

1. H. Chen, S. A. Awadalla, J. Mackenzie, R. Redden, G. Bindley, A. E. Bolotnikov, G. S. Camarda, G. Carini, and R. B. James, IEEE Trans. Nucl. Sci, **54**(4), 811-816 (2007).
2. C. Szeles, phys. stat. sol. (b), **341**(3), 783-790 (2004).

3. M. Funaki, T. Ozaki, K. Satoh, R. Ohno, Nucl. Instrum. Meth. A, **436**, 120-126 (1999).
4. Y. Cui, M. Groza, G. W. Wright, U. N. Roy, A. Burger, L. Li, F. Lu, M. A. Black, and R. B. James, J. Electron. Mater., 35(6), 1267-1274 (2006).
5. K. C. Mandal, S. H. Kang, M. Choi, A. Kargar, M. J. Harrison, D. S. McGregor, A. E. Bolotnikov, G. A. Carini, G. C. Camarda, and R. B. James, IEEE Trans. Nucl. Sci., **54**(4), 802-806 (2007).
6. R. B. James, J. Electron. Mater., **27**, 788-799 (1998).
7. A. Burger, M. Groza, Y. Cui, D. Hillman, E. Brewer, A. Bilikiss, G. W. Wright, L. Li, F. Lu, and R. B. James, J. Electron. Mater., **32**, 756-760 (2003).
8. M. Niraula, K. Yasuda, Y. Nakanishi, K. Uchida, T. Mabuchi, Y. Agata, and K. Suzuki, J. Electron. Mater., 33(6), 645-650 (2004).
9. M. Niraula, K. Yasuda, K. Uchida, Y. Nakanishi, T. Mabuchi, Y. Agata, and K. Suzuki, IEEE Electron. Device Lett., 26(1), 8-10 (2005).
10. K. Yasuda, M. Niraula, H. Kusama, Y. Yamamoto, M. Tominaga, K. Takagi, Y. Agata, and K. Suzuki, IEEE Trans. Nucl. Sci., 52(5), 1951-1955 (2005).
11. M. Niraula, K. Yasuda, H. Ohnishi, H. Takahashi, K. Eguchi, K. Noda, and Y. Agata, J. Cryst. Growth, **284**, 15-19 (2005).
12. K. Yasuda, H. Hatano, T. Ferid, M. Minamide, T. Maejima, and K. Kawamoto, J. Crys. Growth **166**, 612 (1996).
13. M. Ekawa, K. Yasuda, S. Sone, Y. Sugiura, M. Saji, and A. Tanaka, J. Appl. Phys. **67**, 6865 (1990).
14. S. Sone, M. Ekawa, K. Yasuda, Y. Sugiura, M. Saji, and A. Tanaka, Appl. Phys. Lett. **56**, 539 (1990).
15. M. Niraula, K. Yasuda, K. Noda, K. Nakamura, I. Shingu, M. Yokota, M. Omura, S. Minoura, H. Ohashi, R. Tanaka, and Y. Agata, IEEE Trans. Nucl. Sci., **54**(4) 817-820 (2007).
16. M. Niraula, K. Yasuda, K. Takagi, H. Kusama, M. Tominaga, Y. Yamamoto, Y. Agata, and K. Suzuki, J. Electron. Mater., **34** (6), 815-819 (2005).
17. N. Lovergine, M. Bayhan, P. Prete, A. Cola, L. Tapfer, A. M. Mancini, J. Crys. Growth **214/215**, 229 (2000).
18. V. Babentsov, V. Corregidor, J. L. Castano, M. Fiederle, T. Feltgen, K. W. Benz and E. Dieguez, Cryst. Res. Technol., 36(6), 535(2001).
19. M. Niraula, K. Yasuda, A. Watanabe, Y. Kai, H. Ichihashi, W. Yamada, H. Oka, T. Yoneyama, H. Nakashima, T. Nakanishi, K. Matsumoto, D. Katoh, and Y. Agata, IEEE Trans. Nucl. Sci., (2009) (in press).
20. D. P. Halliday, M. D. G. Potter, J. T. Mullins, and A. W. Brinkman, J. Cryst. Growth, **220**, 30, (2000).
21. H. Y. Shin, Y. Sun, Mater. Sci. Eng. B, **52**, 78 (1998).
22. M. Yokota, K. Yasuda, M. Niraula, K. Nakamura, H. Ohashi, R. Tanaka, M. Omura, S. Minoura, I. Shnigu, and Y. Agata, J. Electron. Mater., 37 (9), 1391 (2008).
23. S. M. Sze in "*Semiconductor Devices Physics and Technology*", John Wiley and Sons, pp. 48-55 and 92-97 (1985).

Mater. Res. Soc. Symp. Proc. Vol. 1164 © 2009 Materials Research Society 1164-L05-02

Modeling the Crystal Growth of Cadmium Zinc Telluride: Accomplishments and Future Challenges

Jeffrey J. Derby, David Gasperino, Nan Zhang, and Andrew Yeckel
Department of Chemical Engineering and Materials Science
University of Minnesota
Minneapolis, MN 55455-0132, U.S.A.

ABSTRACT

The availability of large, single crystals of cadmium zinc telluride (CZT) with uniform properties would lead to improved performance of gamma radiation detectors fabricated from them. However, even though CZT crystals are the central element of these systems, there remains relatively little fundamental understanding about how these crystals grow and, especially, how crystal growth conditions affect the properties of grown crystals. This paper discusses the many challenges of growing better CZT crystals and how modeling may favorably impact these challenges. Our thesis is that crystal growth modeling is a powerful tool to complement experiments and characterization. It provides an important approach to close the loop between materials discovery, device research, systems performance, and producibility. Specifically, we discuss our efforts to model gradient freeze furnaces used to grow large CZT crystals at Pacific Northwest National Laboratories and Washington State University. Model results are compared with experimental measurements, and the insight gained from modeling is discussed.

INTRODUCTION

Large, single crystals of cadmium zinc telluride (CZT) form the heart of several advanced gamma detectors, which promise portable, low-cost, and sensitive devices to monitor radioactive materials [1-6]. Decades of development have produced great strides in improved crystal growth processes and better materials for these systems [7,8], and CZT crystals of sufficient quality are now commercially available for simple counting and monitoring applications. However, today's homeland security needs demand large field-of-view imaging and high-sensitivity, high-resolution spectroscopic analysis, which require large, single CZT crystals with spatially uniform charge-transport properties [8]. Affordable material of this size and quality is not yet available.

The growth of large CZT crystals is not well understood and surprisingly less mature than semiconductor crystal growth employed for the electronics industry. One reason for this state of affairs is that CZT crystal growth is far more challenging than that of more traditional semiconductor crystals, such as silicon and gallium arsenide. Indeed, the growth of large, single crystals of CdTe or CZT is notoriously difficult; Rudolph [9,10] details the many challenges encountered during growth. The end result for the growth of CZT is that typical yields of useable material from a crystalline boule remain at 10% or lower [11], resulting in very high materials costs. To improve existing radiation detector crystals or to develop new materials, the performance-property-processing loop must be closed. Namely, device performance must be understood in terms of a mechanistic understanding of crystal growth, and this understanding must be put into the practice of crystal growth using new approaches and modern ideas.

Toward the goal of obtaining a more complete understanding of crystal growth, we are developing and applying computational models of crystal growth processes. Relatively few prior modeling studies have addressed the vertical Bridgman growth of CdTe and CZT. Sen *et al.* [12], Pfeiffer and Mühlberg [13], and Parfeniuk *et al.* [14], employed models that neglected melt convection; however, because of the relatively large influence of convection on heat transfer in this system, this is a poor assumption. We have developed more detailed two-dimensional models to study the application of both vertical [15-18] and horizontal [19-21] Bridgman methods for the production of CZT infrared detector substrates. We followed this work with analyses of high-pressure growth system for CZT radiation detector crystals [22-24]. Yeckel *et al.* [25] studied the three-dimensional effects on melt convection and solute segregation caused by system imperfections in CZT crystal growth in a vertical Bridgman system but employed an idealized representation of furnace heat transfer. Recently, we have used a coupled modeling approach to analyze an industrial system, used by eV Products, Inc. This modeling approach employs a global-scale model for furnace heat transfer coupled to a local-scale model for heat transfer, melt flow, and solidification within the ampoule [26]. Such an approach is needed to relate model predictions to actual process operation. With this approach, we have reported on a successful validation of our model with experimental measurements [27]. In more recent work, we have investigated the feasibility of using furnace temperature profile manipulation in an EDG furnace to control interface shape during growth of CZT [28,29].

Ideally, single-crystals of CZT should be grown in a manner that produces material of uniform chemical composition and with few crystalline defects. Thus far, however, achieving this goal at growth rates suitable for industrial production has been elusive. The application of conventional crystal growth technology to the growth of CZT has suffered from problems of multiple grains, extended defects (primarily dislocations and twins), the formation of second-phase particles enriched with tellurium, and poor reproducibility (synonymous with low yields). Gaining a clear understanding of why these issues arise and how they may be addressed is the overriding objective of the research proposed here.

METHODS AND APPROACH

For the results presented here, we employ the crystal growth simulation software CrysMAS, developed by the Crystal Growth Laboratory of the Fraunhofer Institute of Integrated Systems and Device Technology (IISB) in Erlangen, Germany [30,31]. This package is capable of predicting high temperature heat transfer within complex crystal growth furnaces by solving the energy conservation equations using the finite volume method on an unstructured triangular grid. The radiative heat transfer calculation is implemented using view factors with an enclosure method. CrysMAS applies a structured grid to perform the heat transfer, fluid flow, and phase change computations needed within the ampoule. Furnace setpoint temperatures are specified, and the heater powers are solved as unknowns. CrysMAS employs a quasi-Newton iterative method to arrive at a converged solution.

Our model is based on a Mellen Company electrodynamic gradient freeze (EDG) furnace with 18 controlled heating zones. The axial symmetry of the experimental furnace allows for a simplified two-dimensional model representation in cylindrical coordinate space. Computations are performed specifically for the design and operation of furnaces employed by the groups of Professor Kelvin Lynn of Washington State University (WSU) and Dr. Mary Bliss of Pacific Northwest National Laboratories (PNNL). We report on model development, validation, and more results in [32,33]. A schematic diagram of the computational domain for the model and a

46

representation of the finite volume mesh are shown Fig. 1.

Several meshes were constructed and used to assess numerical convergence and solution accuracy. Accurate computations were achieved with approximately 115,000 degrees of freedom for the PBN ampoule system and 170,000 degrees of freedom for the graphite ampoule system. Solution times for these systems varied, respectively, from approximately 45 seconds per iteration to approximately 140 seconds per iteration on a Dell Precision Workstation outfitted with two Quad Core Intel Xeon Processors running at 2.66GHz.

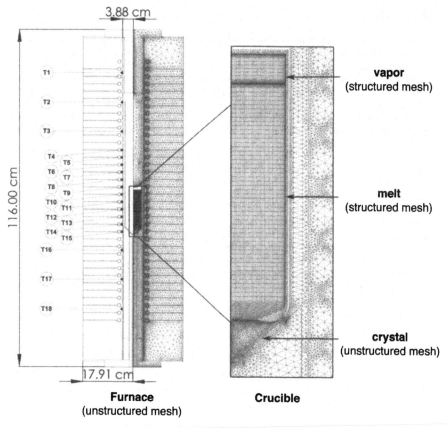

Figure 1: Schematic diagram of crystal growth model (showing the case of the PBN ampoule) constructed using CrysMAS.

RESULTS

We present a series of results that lend insight to strategies that may be employed to address the growth issues discussed previously, namely multiple grains, extended defects, second-phase tellurium particles, and poor reproducibility. In the computations presented here, we compare growth conditions attained by replacing the existing graphite ampoule in the EDG CZT growth system with one fabricated from pyrolitic boron nitride (PBN). Details are provided in [32,33].

Multiple grains

There is relatively little fundamental understanding of grain formation and growth during the crystal growth of CZT. Significantly, attempts using seeded growth of CZT in Bridgman systems have typically been abandoned, and unseeded growth naturally results in the possibility of multiple grains that from simultaneous nucleation events. We argued in [32] that thermal conditions that encourage larger gradients during the undercooling of the melt that precedes nucleation may spatially focus nucleation events. This may discourage the nucleation of many different grains and improve crystallinity. There have also been contentions that kinetic effects, such as forcing fast growth conditions early after nucleation may promote just a few dominant grains via growth competition [17,21].

Grains are also speculated to arise from deleterious interactions between the growth interface and ampoule wall. Figure 2 shows an interesting difference between the EDG growth systems employing a graphite crucible and a PBN crucible. An important difference between the two systems is the shape of the solid-liquid interface. Using simple geometrical arguments, the motion of an interface must always be normal to its shape. Thus a solid-liquid interface shape concave with respect to the crystal will propagate deleterious events, such as the formation of a new grain, inward toward the bulk of the crystal. Conversely, a convex interface will tend to force grains to grow outward. Hence, our model results indicate that the PBN crucible may have an advantage over the graphite crucible, at least in suppressing grain growth early on. Interestingly, Funaki et al. presented evidence that a concave interface shape may lead to multiple grains even if the concave part occurs away from the ampoule wall during the growth of CdTe in a traveling heater method [34].

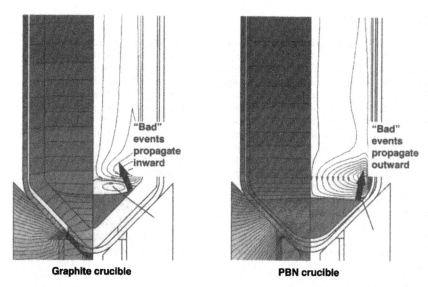

Graphite crucible   **PBN crucible**

Figure 2: Comparison of the state of each EDG CZT system during initial stages of growth. All system parameters are held constant except for the crucible design. Temperature contours are plotted with a spacing of 2 K on the left of each image, and streamfunction contours are spaced at 7 cm²/s on the right. The PBN system exhibits a more favorable interface shape.

Extended defects

Thermal gradients and wall interactions have a profound effect on setting dislocation levels in materials, such as compound semiconductors, that are mechanically weak enough to deform under thermoelastic stresses experienced during growth and cooldown. Indeed, CZT is one of the weakest of compound semiconductors. Thus, the normal prescription to reduce defects is to simply grown under thermal gradients as small as possible. Of course, morphological stability considerations will limit how low gradients can go before stable growth ceases.

If extended defect levels must be lowered significantly, a more dramatic approach may be needed. Of great interest is the detached Bridgman approach, whereby growing crystal in a Bridgman configuration is induced to separate from the ampoule wall [35,36]. Eliminating adverse wall interactions (and stresses) has been shown to reduce dislocation densities by several orders of magnitude [37,38]. Unfortunately, detachment is sporadic and poorly understood. Recently, we have started to investigate the stability of this method [39,40] and hope that such understanding may help advance the growth of CZT by this very promising technique.

Second-phase particles

One of the most vexing issues for all compound materials is the formation of second-phase particles that can markedly degrade the properties of the crystalline material. The source of these particles is most likely thermodynamic in origin. Following the excellent overview by Rudolph [41], a perfectly ordered crystalline array of a compound AB can exist only if the elements A and B are present in exact stoichiometric ratio. Any deviation from stoichiometry must result in the disorder of atoms or in vacancies. To make matters more complicated, at all temperatures above absolute zero, configurational entropic effects (which act to decrease the Gibbs free energy of the solid) will produce non-zero, equilibrium concentrations of vacancies, self-interstitials, and antisite defects. Therefore, at a given temperature and pressure, there is a region of composition of finite extent over which the compound semiconductor exists. This region marks the boundary of the solidus lines of the compound.

As absolute temperature decreases to zero, the width of the existence region vanishes and perfect AB exists as the equilibrium state. The shrinking of the existence region with decreasing temperature is a generic feature of compound semiconductors and leads to an important phenomenon known as retrograde solubility. Namely, the solubility of the excess component in the solid decreases with temperature, and, upon cooling, there is a thermodynamic driving force for the excess component to form its own phase. Now, consider A to represent Cd and B to represent Te (for the moment ignore the Zn atoms that substitute for Cd in CZT). Under typical melt-growth conditions, CZT is grown from a melt that is slightly tellurium rich, thus leading to excess Te at growth temperatures and CZT that is supersaturated with tellurium upon cooldown. Tellurium-rich, second-phase particles form under the action of this unfavorable thermodynamic condition. What is still puzzling, however, is the very large size of these particles in CZT, seemingly too large to be explained by simple solid-state nucleation growth mechanisms.

This is a phenomenon that might be clarified using modeling approaches, such as the defect dynamics computations that have clarified the formation of defects found in melt-grown silicon [42,43]. Such modeling approaches may provide enough insight to tailor growth conditions to limit the growth of these particles, although they will never be eliminated owing to their thermodynamic origin.

Reproducibility

The better understanding of crystal growth systems provided by modeling will lead to better outcomes. However, we believe that the practice of crystal growth will be significantly improved by applying physics-based, computational process models along with modern process control theory. We are developing and applying tools for model-based control of the growth of CZT crystals via the electrodynamic gradient freeze method. SUch approaches promise the possibility of controlling features such as interface shape for improved crystallinity. A schematic representing these ideas is presented in Figure 3. Such approaches have not been widely applied to crystal growth systems growth, and significant increases are expected for crystal quality and yield. We hope to provide results on our efforts in the near future.

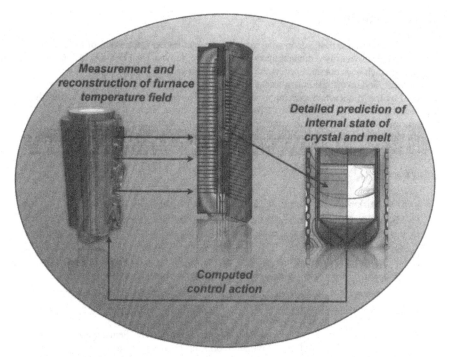

Figure 3: Model-based, feedback control may be the key for the reproducible of CZT crystals. Measurements during growth would be used to compute the furnace thermal field, a detailed prediction of the internal state of the growing crystal would be computed, and an optimal control action would be determined in real time.

FINAL REMARKS

We believe that progress in crystal growth will be accelerated by the careful application of computational models. Clarifying the conditions of crystal growth and assessing the impact of process changes will promote rational decisions for process improvement. Indeed, computations such as performed here are able to predict the impact of design changes far more cheaply and quickly than experiments.

However, much work remains. The fundamental links between macroscopic crystal growth conditions and crystal structure and properties are still poorly understood, especially with regard to microstructural properties of CZT, such as second-phase (tellurium) particle formation. More fundamental studies of the materials science of CZT crystal growth are needed. In addition, the development of advanced, real-time, model-based control methods will be needed to improve the quality and reduce the costs associated with detector-grade CZT production.

ACKNOWLEDGMENTS

This material is based upon work supported in part by the Department of Energy, National Nuclear Security Administration, under Award Number DE-FG52-06NA27498, the content of which does not necessarily reflect the position or policy of the United States Government, and no official endorsement should be inferred. Computational resources were provided by the Minnesota Supercomputing Institute. The authors would like to acknowledge the significant input to this work of Mary Bliss (Pacific Northwest National Laboratory), Kelly Johnson and Kelvin Lynn (Washington State University) and ongoing collaboration on crystal growth modeling software with Jochen Friedrich, Georg Müller, and Thomas Jung (Crystal Growth Laboratory, Fraunhofer Institute IISB).

REFERENCES

1. 1. E. Raiskin and J.F. Butler. *IEEE Trans. Nuclear Science*, 35:81, 1988.
2. J.F. Butler, C.L. Lingren, and F.P. Doty. *IEEE Trans. Nucl. Phys.*, 39:605, 1992.
3. F.P. Doty, J.F. Butler, J.F. Schetzina, and K.A. Bowers. *J. Vac. Sci. Technol. B*, 10:1418, 1992.
4. J.F. Butler, B. Apotovsky, A. Niemela, and H. Sipila. In *Proceedings of the SPIE*, volume 2009, page p. 121. SPIE, Bellingham, WA, 1993.
5. R.B. James, T.E. Schlesinger, J. Lund, and M. Schieber. In T.E. Schlesinger and R.B. James, editors, *Semiconductors for Room Temperature Nuclear Detector Applications*, volume 43, page p. 335. Academic Press, San Diego, 1995.
6. R.B. James and P. Siffert, editors, "Room Temperature Semiconductor Detectors: Proceedings of the 11th International Workshop on Room Temperature Semiconductor X- and Gamma-Ray Detectors and Associated Electronics,"*Nuclear Instruments and Methods in Physics Research A*, volume 458, 2001.
7. Szeles, C. Cameron, S.E. Ndap, J.-O. Chalmers, W.C. *IEEE Trans. Nuclear Science*, 49:2535, 2002.
8. C. Szeles, S.E. Cameron, S. A. Soldner, J.-O. Ndap, And M. D. Reed, *Journal of ELECTRONIC MATERIALS* Vol.33, 742–752 (2004).
9. P. Rudolph, Progr. Crystal Growth and Charact. 29, 275 (1994).
10. P. Rudolph. In M. Isshiki, editor, *Recent Development of Bulk Crystal Growth*. Research Signpost, Trivandrum, India, 1998.
11. J.J. Griesmer, B. Kline, J. Grosholz, K. Parnham, D. Gagnon, In: Proceedings of IEEE MIC 2001, San Diego, Nov. 2001.
12. S. Sen, W.H. Konkel, S.J. Tighe, L.G. Bland, S.R. Sharma, and R.E. Taylor. *J. Crystal Growth*, 86:111-117, 1988.
13. M. Pfeiffer and M. Mühlberg. *J. Crystal Growth*, 118:269, 1992.
14. C. Parfeniuk, F. Weinberg, I.V. Samarasekera, C. Schvezov, and L. Li. *J. Crystal Growth*, 119:261, 1992.
15. S. Kuppurao, S. Brandon, and J.J. Derby, *J. Crystal Growth* **155**, 93–102 (1995).
16. S. Kuppurao, S. Brandon, and J.J. Derby, *J. Crystal Growth* **155**, 103–111 (1995).

17. S. Kuppurao, S. Brandon, and J.J. Derby, *J. Crystal Growth* **158**, 459–470 (1996).

18. S. Kuppurao and J.J. Derby, *J. Crystal Growth* **172**, 350–360 (1997).

19. K. Edwards and J.J. Derby. *J. Crystal Growth*, 179, 120, 1997.

20. K. Edwards and J.J. Derby. *J. Crystal Growth*, 179, 133, 1997.

21. K. Edwards and J.J. Derby. *J. Crystal Growth*, 206, 37–50, 1999.

22. A. Yeckel, F.P. Doty, and J.J. Derby, J. Crystal Growth 203, 87–102 (1999).

23. A. Yeckel and J.J. Derby, J. Crystal Growth 209, 734–750 (2000).

24. A. Yeckel and J.J. Derby, J. Crystal Growth 233, 599–608 (2001).

25. A. Yeckel, G. Compere, A. Pandy, and J.J. Derby. *J. Crystal Growth*, 263:629–644, 2004.

26. A. Yeckel, A. Pandy, and J.J. Derby, *Int. J. Numer. Meth. Engng.* **67**, 1768–1789 (2006).

27. Pandy, A. Yeckel, M. Reed, C. Szeles, M. Hainke, G. Müller, and J.J. Derby, *J. Crystal Growth* **276**, 133–147 (2005).

28. L. Lun, A. Yeckel, C. Szeles, M. Reed, P. Daoutidis, and J.J. Derby, *J. Crystal Growth* **290**, 35–43 (2006).

29. L. Lun, A. Yeckel, J.J. Derby, and P. Daoutidis, in: *Proceedings of the IEEE 2007 Mediterranean Conference on Control and Automation (MED 2007),* Athens, Greece, June 27–29, 2007.

30. M. Kurz, A. Pusztai, and G. Müller, *J. Crystal Growth*, 198:101, 1999.

31. R. Backofen, M. Kurz, and G. Müller, *J. Crystal Growth*, 199:210, 2000.

32. D. Gasperino, K. Jones, K. Lynn, M. Bliss, and J.J. Derby, *J. Crystal Growth*, to be submitted, 2008.

33. D. Gasperino, Ph.D. thesis, University of Minnesota, in preparation.

34. M. Funaki, H. Shiraki, M. Tamaki, Y. Mito, and R. Ohno, Presentation L3.2 at the 2009 Spring MRS Meeting.

35. T. Duffar, I. Paret-Harter, and P. Dusserre, J. Crystal Growth 100 (1990) 171–184.

36. L.L. Regel and W.R. Wilcox, Microgravity Sci. Technol. 14 (1999) 152–166.

37. D.J. Larson, Microgravity News 1(6) (Winter 1994) 10.

38. T. Duffar, P. Dusserre, N. Giacometti, K.W. Benz, M. Fiederle, J.C. Launay, E. Dieguez, and G. Roosen, Acta Astronautica 48 (2001) 157–161.

39. C. Stelian, A. Yeckel, and J.J. Derby, J. Crystal Growth 311, 2572-2579 (2009).

40. C. Stelian, M.P. Volz, and J.J. Derby, J. Crystal Growth in press (2009).

41. P. Rudolph. Cryst. Res. Technol. 38:542, 2004.

42. T. Sinno, J. Crystal Growth 303, 5–11 (2007)

43. T. Sinno, E. Dornberger, W. von Ammon, R.A. Brown, and F. Dupret, Materials Science and Engineering, 28 (2000) 149-198.

Neutron Detectors

Mater. Res. Soc. Symp. Proc. Vol. 1164 © 2009 Materials Research Society

Improved Fabrication Technique for Microstructured Solid-State Neutron Detectors

S.L. Bellinger, W.J. McNeil, D.S. McGregor

S.M.A.R.T. Laboratory, Mechanical and Nuclear Engineering Dept., Kansas State University, Manhattan, KS 66506, U.S.A.

ABSTRACT

Microstructured semiconductor neutron detectors have superior efficiency performance over thin-film coated planar semiconductor detectors. The microstructured detectors have patterns deeply etched into the semiconductor substrates subsequently backfilled with neutron reactive materials. The detectors operate as pn junction diodes. Two variations of the diodes have been fabricated, which either have a rectifying pn junction selectively formed around the etched microstructures or have pn junctions conformally diffused inside the microstructures. The devices with the pn junctions formed in the perforations have lower leakage currents and better signal formation than the devices with selective pn junctions around the etched patterns. Further, pulse height spectra from conformally diffused detectors have the main features predicted by theoretical models, whereas pulse height spectra from the selectively diffused detectors generally do not show these features. The improved performance of the conformal devices is attributed to stronger and more uniform electric fields in the detector active region. Also, system noise, which is directly related to leakage current, has been dramatically reduced as a result of the conformal diffusion fabrication technique. A sinusoidal patterned device with 100 μm deep perforations backfilled with ^{6}LiF was determined to have 11.9 ± 0.078% intrinsic detection efficiency for 0.0253 eV neutrons, as calibrated with thin-film planar semiconductor devices and a ^{3}He proportional counter.

INTRODUCTION

Compact neutron detectors with high counting efficiency and low voltage operation can be fabricated by etching micro-cavity patterns into a semiconductor diode [1]. When these microscopic patterns are filled with a neutron conversion material such as ^{10}B or ^{6}LiF, the energetic reaction products in the form of charge particles can be captured within a silicon diode. The deep trenches backfilled with neutron reactive material increase the neutron absorption efficiency, and for devices with narrow trenches, the probability of registering the energetic reaction products is increased [1-3]. Overall, the neutron detection efficiency is significantly increased over that of a thin-film neutron detector [3-8]. Construction of the detectors has many challenges, which include minimizing the leakage current while achieving a good signal to noise ratio.

Deeply etched structures are easily fabricated with inductively-coupled plasma etching technology (Fig. 1) [4,6,9]. However, leakage current and noise are serious issues when introducing the deep vertical structures into a semiconductor device, such as a pn junction diode radiation detector. Leakage current contributes to electronic noise and reduces the ability to detect small signals. Hence, much work has been done to minimize the leakage current. A series of silicon fabrication techniques has been reported elsewhere that can reduce leakage current by orders of magnitude [3]. For instance, selective diode patterns with a separation distance between the diode and the etched sidewalls of the trenches have shown substantial improvements as well as the addition of an oxide passivation in the trenches. Yet, greater improvement is reported in the present work by fabricating the diode structure within the perforation (trenches in the present case). Similar to porous silicon diode devices, the diode conforms all around the sidewalls and bottom of the deep trench structures [10]. The resulting irregular-shaped diode becomes useful for neutron detection when the trenches are filled with neutron conversion material.

Figure 1. Microstructure of the perforated patterns etched into a Si substrate, showing detail of the sinusoidal pattern.

THEORY

Fabrication processes for deep microstructured devices were developed to reduce surface damage and contamination from processing techniques. Plasma-etched surfaces are a source of crystalline damage and possible contamination, which can reduce charge carrier lifetimes and increase recombination rates for charge carriers near the sides and the bottoms of the perforations. Selectively diffused devices have perforations etched into the semiconductor surface with the *pn* junction rectifying contact fabricated between the etched perforations, as shown in Fig. 2. Damaged surfaces cause increased leakage current and detector noise. It has been shown that the leakage current can be significantly reduced by growing a thermal oxide along the sidewalls and bottoms of the perforations [3] , yet the pulse height spectra indicate that the depletion depths do not fully extend past the perforation depths [4], thereby reducing the measured efficiency.

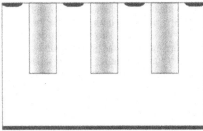

Figure 2. Perforated diode with selective diffused configuration.

The conformal design uses a rectifying *pn* junction diffused directly into the perforations. This different approach negates the impact of surface damage and contamination by saturating the surface with high concentrations of controlled dopants. Furthermore, the electric field distribution is changed from the selective diffused or planar design, such that a high electric field and the depletion region are both extended to the bottom of the perforations; hence the diode conforms to the surfaces of the etched structures (Fig. 3).

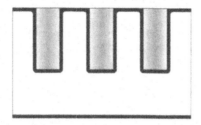

Figure 3. Perforated diode with conformal diffused configuration.

Fig. 4 (left) shows the potential distribution for a selectively diffused detector, in which the potential steadily diminishes through the bulk of the structure. The equipotential lines expand further apart at deeper locations in the device. Thus, the electric field is decreasing towards the bottom of the trench, which has an adverse effect on the selective diode configuration because good signal formation depends on both swift charge transport within a high electric field and a low recombination rate within the depletion region. With the selective diode design, it is necessary to apply reverse bias to extend the depletion region just beyond the depth of the trenches, even with very high resistivity material. Further, the magnitude of the electric field at the bottom of the trench may be significantly reduced below that near the top, and will produce much smaller signal pulses for all charged particle tracks occurring near the trench bottom.

The conformal diode is fabricated with heavy doping within the trench structure. Shown in Fig. 4 (right), the potential remains uniform along the entire surface of the trench. For self-bias, where no external bias is applied to the device, the potential drops evenly away from all surfaces of the trench toward the bulk, indicating that the electric field is relatively uniform and higher in magnitude along the trench than the selectively diffused design. Hence, regardless of trench depth, charge carriers are born in a region of high electric field and promptly drifted in the active region of a conformal diffused diode.

The selective diffused microstructured diode design is prone to have smaller signal pulses from charge carriers excited in the deep locations along the trench. The lower pulse heights are due to a reduced electric field, and possibly combined with a truncated depletion region, thereby increasing charge carrier recombination. Signal pulses that are small in magnitude will shift toward the noise floor of the pulse height spectrum, blending with electronic noise. Further, if the depletion region is shorter than the trench depths, the maximum theoretical neutron detection efficiency is compromised. However, the conformal diode design extends the depletion region and strong electric field beyond the full depth of the trench, thereby allowing a signal to be formed completely from the top surface to the full depth of the trench. As a result, the pulses will be larger and will appear in higher energy channels well above the noise floor, and the full effect of the perforations on neutron detection efficiency should be realized.

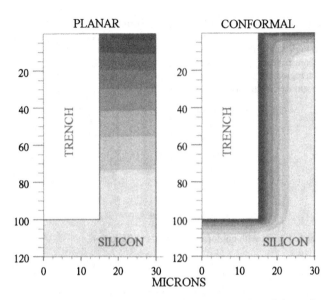

Figure 4. Potential maps of the etched selective diffused diode (left) and the conformal diffused diode (right) configurations.

EXPERIMENTAL

A sinusoidal pattern for the trench structure is an effective geometry for uniform neutron detection [5]. Thus, the conformal diode configuration was fabricated in sinusoidal structures. Three tests were performed for the conformal configuration and compared to identical devices fabricated with the selective diffused diode configuration. These comparative tests included reverse bias leakage current, reverse bias capacitance, and thermal neutron detection efficiency.

Device Fabrication

Compared to previous perforated diode designs using selective diffusion and oxide passivation, the conformal diode fabrication process has fewer process and cleaning steps, and a relaxed thermal processing budget. The diode fabrication processes were performed on 7.6 cm diameter float zone, single-side polished, >10 kΩ-cm, n-type Si wafers approximately 325 μm thick. The raw wafers were dehydrated at 150°C in a clean room oven for 30 minutes and subsequently treated with HMDS primer to promote photoresist adhesion. AZ nLof 2070 photoresist was applied at 2000rpm for 30 seconds to attain an 8.25μm thick photoresist layer. The sinusoidal masking pattern was then exposed and subsequently developed to form the dry etching mask as shown in Fig. 5.

The wafers were dry etched with the HARDE ICP-RIE etch process [9]. The residual photoresist after dry etching was removed by plasma ashing the wafers in a pure oxygen plasma for 2 hours. Afterwards, the wafers were chemically cleaned with a Piranha bath at 130°C for 15 minutes, followed by a Baker Clean® bath at 70°C for 15 minutes. The wafers were then BOE etched for 20 seconds and immediately spin-rinsed and placed in an oxidation furnace. A 2 hour wet oxide was grown on the wafers at 1150°C to create a 0.8μm thick thermal oxide layer to provide a diffusion mask and a field oxide around the device after fabrication.

The wafers were again patterned with photoresist, using an in-house recipe developed for photoresist-bridging (see Fig. 5), to produce 6mm diameter diffusion windows over the pre-existing trenches. Once the photoresist pattern was developed, the wafer was chemically etched in a 6:1 buffered oxide etch (BOE) solution to remove the silicon dioxide in the pre-described diffusion areas. The masking photoresist was then removed with Kwik Strip® and the wafers were again cleaned with the Piranha and Baker Clean® baths. The wafers were BOE etched for 20 seconds and immediately spin-rinsed and placed in a boron nitride solid-source diffusion furnace. A 2.5 hour diffusion soak was performed on the wafers at 950°C to create a shallow p-type junction within the microstructured perforated trenches of the n-type substrate material.

After the diffusion process, the wafers were then dipped in BOE for 2 minutes to remove the surface oxides. The wafers were patterned with photoresist, again using an in-house recipe developed for photoresist-bridging and liftoff, for the 6mm O.D. x 5.75mm I.D. metal ring contacts. Once the photoresist pattern was developed, the wafer was chemically etched in a BOE solution to remove any residual silicon oxides in the metal contact pattern. The wafers were immediately placed in the chamber of an e-beam evaporation system and the chamber was pumped down to a vacuum of $2x10^{-6}$ Torr. Two layers of metal were evaporated on the wafer consisting of a 500Å Ti layer followed by a 2500Å Al layer. Afterwards, the wafer backside was also coated with a 500Å Ti layer followed by a 2500Å Al layer.

The liftoff photoresist and metal were then removed with Kwik Strip®. The perforations were backfilled with ^{6}LiF powder material and subsequently coated with an e-beam evaporated 26 μm layer of ^{6}LiF through an indexed shadow mask. The wafer was then selectively coated, using the same indexed shadow mask, with a protective coating of Humiseal®. Afterwards, the wafer was diced into individual 7mm x 7mm detector die, and each detector was mounted for testing on a 4 pin header with a metal cap for EM shielding.

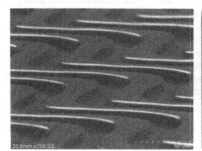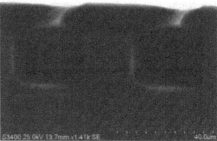

Figure 5. Applied dry-etching photoresist mask on bare silicon (left). Photoresist bridging, used to pattern selective removal of thermally grown silicon oxide, for conformal diffusion into the trenches (right).

Leakage Current Measurement

Rectifying junction leakage current characteristics were measured using a Keithley Instruments IV-curve-tracing system and an EM shielded dark box. The reverse bias leakage current (or dark current) for a 100 μm deep sinusoidal conformal diffused diode was compared to that of a selective diffused junction device (with identical perforation dimensions and depths) and a simple thin-film planar device [3].

Capacitance Measurement

Rectifying junction capacitance characteristics were measured using a Hewlett-Packard CV-curve-tracing 1MHz system and an EM shielded dark box. The reverse bias capacitance characteristics for the conformal perforated device, selectively diffused device, and a planar diffused device were measured and compared.

Thermal Neutron Counting Efficiency

A diffracted neutron beam from the Kansas State University TRIGA Mark II nuclear reactor was used for the efficiency calibrations. The beam diffraction angle was set to yield 0.0253 eV thermalized neutrons to the work space. The neutron beam was stopped down to a diameter of 1.27 cm with a cadmium shutter. The ^{3}He proportional counter used was a 2-inch diameter, 6-inch long Reuter Stokes model P4-1603-207. Details of the calibration method can be found elsewhere [11].

The neutron beam flux was measured before and after each neutron detector response experiment to guarantee consistency in the neutron flux. A cadmium shutter was used to distinguish the thermal neutron response from fast and epithermal neutron response. The detector was tested with a reverse bias of 0 volts and 10 volts. A standard Ortec 142 preamplifier and NIM counting electronics along with an MCA were used for data collection. The active detection area for the perforated neutron detectors was 0.28 cm^2.

For these efficiency tests, the detector face was orientated perpendicular to the neutron beam. The lower level discriminator (LLD) was chosen such gamma-ray interactions would not register. A distribution of gamma-ray counts can usually be observed at very low signal levels near the noise floor in these devices [4]. Prompt gamma-ray response from neutron activation in the cadmium shutter was used to set the LLD. The measured neutron detection response was then divided by the ^{3}He calibrated neutron flux to determine the thermal neutron detection efficiency.

RESULTS

Conformal diode devices show low leakage currents compared to that of the selective diffused diode design. The capacitance compares well to theoretical predictions and also to the fabricated simple planar devices. The thermal neutron counting efficiency of the conformal device showed improvement in the definition of the spectral features and the signal-to-noise ratio as compared to previous designs [4,5,11]. The ^{3}He calibrated thermal neutron flux was verified by comparing the neutron detection efficiency of a calibrated thin-film ^{6}LiF planar device.

Leakage Current and Device Capacitance

Leakage currents for the various devices are shown in Fig. 6. Notice that the conformal device leakage current at a reverse bias of 10 volts is almost 2 orders of magnitude lower than the leakage current at the same voltage for the selective diffused etched device. Each of these devices has an area of 0.28 cm^2. The traditional simple planar device has the lowest leakage current, yet the conformal device has low enough leakage current such that it is not an impediment to neutron counting.

Capacitance plots for the varying device designs are shown in Fig. 7. Notice that the conformal device capacitance is comparable to the selective diffused device. Both micro-structured devices have lower capacitance than the traditional simple planar device, an unexpected result, which may be due to the reduced Si detector volume from the perforations.

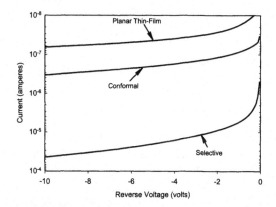

Figure 6. Leakage current from a planar diffused detector, a selective diffused detector with 100 μm deep trenches, and a conformal diffused detector with 100 deep trenches. All devices were 6 mm in diameter.

Thermal Neutron Counting Efficiency

The neutron counting efficiency was determined with the LLD set at channel 200 (approximately equivalent to 300 keV) for both selective and conformal diffused detectors in order to eliminate gamma-ray background counts. The intrinsic efficiency for 0.0253 eV neutrons was calculated by comparing the total neutron counts collected from the detector by the neutron flux determined with the calibrated standard ^{3}He detector. The resulting efficiencies were 8.96 ± 0.059% and 11.94 ± 0.078% for the selectively diffused design and the conformal diffused design, respectively. The conformal diffused diode detector is collecting more neutron counts for the same size of device and perforation depths than the selectively diffused device. The improvement provides more evidence for an increased depletion depth and larger active region in the conformal diffused device over that of the selective diffused device [4].

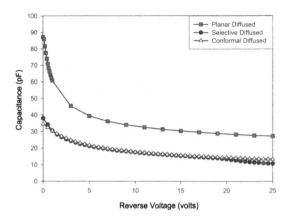

Figure 7. Capacitance plot from a planar diffused, a selective diffused and a conformal diffused detector. The trenches are 100 μm deep for the selective and conformal diffused detectors. All devices were 6 mm in diameter.

CONCLUSIONS

Fabricating a 3D microstructured neutron detector by diffusing a *pn*-junction into the deep structures effectively reduces leakage current and results in better signal response. The improved carrier transport characteristics are reflected in the pulse height spectra. More pulses have a larger magnitude, shifting the spectrum towards higher channels, away from the noise floor thereby providing a better signal-to-noise ratio on average. The result is a more robust detection system with better gamma-ray discrimination, and the device is easier to fabricate than previous designs of microstructured diode detectors. For selective diffused diode configurations, the pulse height distribution resembles an exponential noise spectrum [4]. The non-uniform signal formation in the selective diffused microstructured design can be attributed to charge carriers being excited near and beyond the depletion boundary, thus resulting in little to no signal formation from interactions that deep in the structure. Increased bias and changing depletion should have revealed such an occurrence, but the selective diffused design suffers from excessive leakage current and noise with low to moderate reverse bias, hence could not be operated at higher voltages. The conformal diode design produces a spectral shape that is completely distinguishable from the noise spectra, and has the characteristics features expected from the modeled results [11,12].

ACKNOWLEDGMENTS

This work was supported in part by DRTA contract DTRA-01-03-C-0051, NSF grant no. 0412208, and DOE grant number DE-FG07-04ID14599.

64

REFERENCES

1. J. K. Shultis and D. S. McGregor, *IEEE Trans. Nucl. Sci.* **NS-53**, 1659-1665 (2006).
2. D. S. McGregor, R. T. Klann, H. K. Gersch, E. Ariesanti, J. D. Sanders, and B. VanDerElzen, *IEEE Trans. Nucl. Sci.*, **NS-49**, 1999-2004 (2002).
3. W. J. McNeil, S. L. Bellinger, T. C. Unruh, E. L. Patterson, J. K. Shultis, and D. S. McGregor, (IEEE NSS Conf. Record, San Diego, CA, 2006) pp. 3732-3735.
4. S.L. Bellinger, W.J. McNeil, T.C. Unruh, D.S. McGregor, *IEEE Trans. Nucl. Sci.* (2009) (in press).
5. S.L. Bellinger, W.J. McNeil, T.C. Unruh, D.S. McGregor, (IEEE NSS Conf. Record, Waikiki, HI, 2007) pp. 1904-1907.
6. J. Uher, C. Fröjdh, J. Jakůbek, C. Kenney, Z. Kohout, V. Linhart, S. Parker, S. Petersson, S. Pospíšil, G. Thungström, *Nucl. Instrum. Meth.* **A576**, 32-37 (2007).
7. R. J. Nikolic, A. M. Conway, C. E. Reinhardt, R. T. Graff, T. F. Wang, N. Deo, and C. L. Cheung, (IEEE NSS Conf. Record, Waikiki, HI, 2007), pp. 1577-1580.
8. D.S. McGregor, S.L. Bellinger, D. Bruno, S.A. Cowley, W.L. Dunn, M. Elazegui, A. Kargar, W.J. McNeil, H. Oyenan, E. Patterson, J.K. Shultis, G. Singh, C.J. Solomon, T.C. Unruh, (IEEE NSS Conf. Record, Waikiki, HI, 2007) pp. 808-812.
9. B. Rice, *Inductively Coupled Plasma Etching of Silicon and Gallium Arsenide*, (MS thesis, Kansas State University, 2003).
10. J.P. Clarkson, W. Sun, K. D. Hirschman, L. L. Gadeken, P. M. Fauchet", *Phys. Stat. Sol. (a)* **5**, 1536-1540 (2007).
11. D.S. McGregor, W. J. McNeil, S. L. Bellinger, T. C. Unruh, J. K. Shultis, *Nucl. Instrum. Meth.* (2009) (in review).
12. J.K. Shultis and D.S. McGregor, *Nucl. Instrum. and Meth.* (2009) (in press).

Mater. Res. Soc. Symp. Proc. Vol. 1164 © 2009 Materials Research Society 1164-L06-02

Growth of Boron Carbide Crystals from a Copper Flux

Yi Zhang[1], J.H.Edgar[1], Jack Plummer[1], Clinton Whiteley[1], Hui Chen[2], Yu Zhang[2], Michael Dudley[2], Yinyan Gong[3], James Gray[3] and Martin Kuball[3]
[1]Department of Chemical Engineering, Kansas State University, Manhattan, KS, 66506
[2]Department of Materials Science and Engineering, Stony Brook University, Stony Brook, NY, 11794
[3]H.H.Wills Physics Laboratory, University of Bristol, Bristol, BS8 1TL, United Kingdom

ABSTRACT

Boron carbide crystals ranging in size from 50 microns to several millimeters have been grown from a copper-boron carbide flux at temperatures from 1500°C to 1750°C. The crystal size increased with growth temperature although copper evaporation limited growth at the higher temperatures. Synchrotron X-ray Laue patterns were indexed according to (001) orientation boron carbide structure, indicating the bulk crystals were single crystalline with {001} growth facets. Raman spectrum of boron carbide indicates an improved crystal quality compared to the source powder, but peaks of crystals grown from ^{11}B -enriched source shifted to the lower energy by 1-4 cm^{-1} from literature values, possibly due to the boron isotope dependency. Five fold symmetry defects and twin planes were common as observed by optical microscope and scanning electron microscope. Raindrop shape etch pits were formed after defect selective etching in molten potassium hydroxide at 600°C for 6 minutes. Typically, the etch pit density was on the order of 10^{6} /cm^{2}.

INTRODUCTION

Boron carbide is a good candidate for nuclear applications, such as neutron shielding and neutron detection due to the high neutron absorption capability of ^{10}B, which has thermal neutron absorption cross section of 3873 barn [1]. ^{10}B undergoes the following neutron capture reactions when exposed to neutrons:

$$^{10}B + n \rightarrow Li(0.84MeV) + {}^4He(1.47MeV) + \gamma(0.48MeV)$$
$$^{10}B + n \rightarrow {}^7Li(0.84MeV) + {}^4He(1.47MeV)$$

The resulting large kinetic energies listed in parentheses leads to local heating of the solid. In addition, it is one of the icosahedral boron-riched solids, which have an unusual self-healing ability from radiation damage. Therefore, Emin and Aselage [2] proposed a solid-state boron carbide neutron detector by measuring the Seebeck emf induced by the neutron absorption. Robertson *et al* [3, 4] made a boron carbide-silicon heterojunction diode to detect neutrons by means of collecting the electrons produced by the highly energetic Li and He ions as a result of neutron capture.

The properties of boron carbide are associated with its unique crystal structure. It is based on twelve-boron-atom icosahedra [5], which reside at the corners of an α-rhombohedral unit cell, and three-atom C-C-C chains lying along the rhombohedral [111] axis. Such a complicate structure makes it difficult to produce large, single crystals with low defect densities. Chemical vapor deposition (CVD) methods [6-11] have been extensively studied and developed for boron carbide production. However, for solid-state devices, high quality crystals with good structural order, large size, and low defect densities are necessary. Millimeter size, high quality boron carbide crystals can be produced from a copper-boron carbide solution [12]. In this approach, boron carbide dissolves in the molten copper and forms a solution. As this solution is slowly cooled, boron carbide crystals nucleate and grow. The impact of temperature on the crystal size and quality was determined in this study.

EXPERIMENT

Boron carbide crystals were produced from a saturated copper flux. Natural boron carbide, with 20% B-10 isotope, and B-11 enriched (>99%) boron carbide were used as source material. Boron carbide and copper powder, weight ratio of 1:9, were mixed in a boron nitride crucible. The mixture was held in a high temperature graphite furnace and heated to a temperature between 1500 and 1750°C in 50 °C increments to produce a saturated solution. Argon flowed through the furnace to maintain 1atm pressure inside of the furnace. The solution was held for a period between 1 to 12 hours, and then slowly cooled at 10°C /hour. Crystalline boron carbide embedded in a copper matrix was obtained via this process. The copper was dissolved away by the nitric acid and boron carbide crystals were filtered out of the acid.

To optimize the process conditions, two variables in the process were changed, the maximum temperature and the hold time, the effect of which was evaluated in terms of crystal size and quality. The maximum temperature was varied from 1500 to 1750°C with 50°C increment and the hold time was changed among 1, 2 and 12 hours.

Raman spectroscopy, synchrotron white beam x-ray topography, and defect selective etching were utilized to evaluate the crystal quality. Defect selective etching was carried out to assess the dislocation density via etch pit densities. The boron carbide crystals were placed in the molten potassium hydroxide at 600°C for 6 minutes.

RESULTS

Well-faceted individual crystals ranging in size from 50 microns to several millimeters were produced.

Surface morphology

In most cases, crystals grew in groups with an uneven fracture surface where crystals separated from the group. Hexagonal and prismatic crystals were produced, examples of which are shown in Figure 1. Two typical types of defects were observed, five-fold symmetry defects and twin plane, as shown in Figure 2. Twin planes, where crystals grew along two symmetry orientations from the twin boundary, were common defects existing in boron carbide crystals produced by Aselage [12].

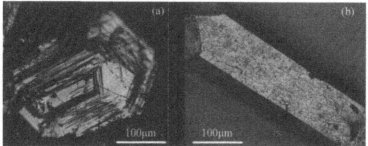

Figure 1. Optical images of boron carbide crystals with (a) hexagonal, (b) prismatic and shape.

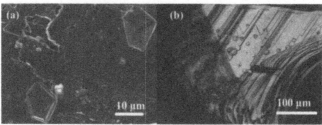

Figure 2. Two typical defects present in carbide crystals. (a) SEM image of five-fold symmetry (b) optical image of twin boundary

<u>**Effect of process conditions**</u>

Higher temperatures, from 1600°C to 1750°C, produced larger crystals. However, the copper evaporation became significant at such high temperatures, which limited the hold time at temperature. An even higher temperature, 1800°C, was tried, but no crystals were produced due to the evaporation of copper. In comparison, at relative low temperatures, from 1500°C to 1600°C, the loss of copper is small, making longer hold time possible. Therefore, more crystals were produced, but the crystal size was smaller.

The largest crystal was obtained at 1750°C and a 2-hour hold. The greatest amount of crystals was produced at 1650 °C and 12 hours hold. Ultimately, the best growth conditions are 1700°C and 12 hours hold time.

<u>**Crystal quality**</u>

The Raman spectrum of two boron carbide crystals grown at different temperatures were displayed Figure 3. The Raman spectra of the crystal grown at 1700°C is consistent with the literature spectra of single crystalline boron carbide with a 4:1 atomic ratio of boron to carbon [13]. The two prominent, narrow bands at 481 and 534 cm^{-1} suggests that the chain structure is well-ordered. However, for the crystals grown from ^{11}B -enriched source, these peaks are shifted to lower energy by 1-4 cm^{-1} compared to those grown from natural boron carbide source, which may be due the boron isotope dependency. The Raman spectrum of [11]B -enriched boron carbide and [10]B -enriched boron carbide recorded by Aselage [14] indicated that [11]B -enriched samples result in a shift to the lower energy.

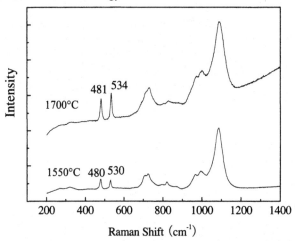

Figure 3. Raman spectrum of boron carbide crystals grown at various temperatures.

Well defined X-ray Laue patterns from synchrotron white beam X-ray topography (SWBXT), shown in Figure 4, were indexed according to (001) orientation boron carbide structure, indicating the bulk crystals were single crystalline with (001) growth planes.

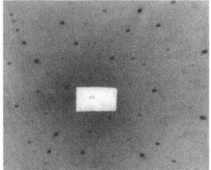

Figure 4. X-ray Laue pattern of boron carbide crystals from SWBXT

Raindrop-shape etch pits were the major shape formed on these crystals. Parallel grooves were also formed (Figure 5 (a)). Because the etch pits are all oriented in the same direction, as clearly displayed in a SEM image, Figure 5 (b), this suggests that they are associated with crystalline defects, possibly dislocations. The etch pits density of the crystal grown at 1700°C is on the order of 10^6 etch pits/cm².

70

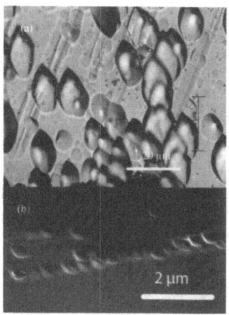

Figure 5. The etch pits in the crystals (a) optical image of raindrop-shape etch pits (b) SEM image, all etch pits oriented in the same direction along a line

CONCLUSIONS

Boron carbide crystals, ranging in sizes from 50 microns to several millimeters, were successfully produced from a copper flux. Higher temperatures, from 1600°C to 1750°C, produced larger crystals. At relative low temperatures, from 1500°C to 1600°C, more crystals were produced, but the crystal size was smaller. Five fold symmetry defects and twin planes were typically observed. The best growth conditions were 1700°C and 12 hours hold time.

The Raman spectrum of boron carbide demonstrated single crystalline boron carbide crystals with 4:1 boron to carbon ratio were produced. However, for the crystals grown from [11]B-enriched source, peaks are shifted by 1-4 cm[-1] from literature values possibly due to the boron isotope dependency. X-ray Laue patterns from synchrotron white beam X-ray topography (SWBXT) indicated that the bulk crystals were single crystalline with {001} growth facets. The typical etch pit density revealed by defect selective etching in these crystals was 10^6 etch pits/cm^2.

ACKNOWLEDGMENTS

This work was supported by National Science Foundation, program of materials world network, grant numbers DMR062807 and DMR0836150, and the United Kingdom Engineering and Physical Science Research Council, grant number EP/D075033/1.

REFERENCES

1. A.O.Sezer and J.I.Brand, Mater. Sci. and Engg. B79, 191 (2001).
2. D.Emin and T.L.Aselage, J. Appl. Phys. 97, 019 (2005).
3. B.W.Robertson, S. Adenwalla, A.Harken, P.Welsch, J.I.Brand, P.A.Dowben, J.P.Classen, Appl. Phys. Lett. 90, 3644(2002)
4. A.N. Caruso, P.A.Bowben, S.Balkir, Nathan Schemm, Kevin Osberg, R.W.Fairchild, Oscar Barrios Flores, Snjezana Balaz, A.D. Harken, B.W. Robertson, and J. I.Brand, Mater. Sci and Engg. B 135 , 129 (2006)
5. D.Emin, Phys. Today 1, 55(1987).
6. J. Winter, H.G. Esser, H. Reimer, L.Grobusch, J. Von Seggern and P. Wienhold, J. Nucl. Mat. 176-177, 486 (1990).
7. K. Shirai and S. Gonda, J. Appl. Phys. 67, 6287 (1990).
8. V.M.Sharapov, A.I.Kanaev, A.P.Zakharov and A.E.Gorodetsky, J. Nucl. Mat. 508, 191(1992).
9. D. Byum, S.Hwang, P.A.Dowben, F.K.Perkins, F.Filips, and N.J. Ianno, Appl. Phys.Lett. 64, 1968(1994).
10. S.V.Deshpande, E.Gulari, S.J. Harris and A.M.Weiner, Appl. Phys. Lett., 65, 1757 (1994).
11. J.C.Oliveira andO. Conde, Thin Solid Films 307, 29 (1997).
12. T.L.Aselage, S.B.Vendeusen and B.Morosin, J. Less-Common Metals 166, 29 (1990).
13. D.R.Tallant, T.L.Aselage, A.N.Campbell, and D.Emin, Phys. Rev. B 40, 5649 (1989).
14. T.L.Aselage, D.R.Tallant and D.Emin, Phys. Rev. B 56, 3124 (1997).

Mater. Res. Soc. Symp. Proc. Vol. 1164 © 2009 Materials Research Society   1164-L06-03

Attempt to Grow α-Rhombohedral Boron Crystals in Copper Solvent

W. Gao[1], C. Whiteley[1], Y. Zhang[1], J. Plummer[1] and J.H. Edgar[1],
[1]Kansas State University, Department of Chemical Engineering,
Manhattan, KS 66506-5102

Y.Y. Gong[2] and M. Kuball
[2]H.H. Willis Physics Laboratory, University of Bristol,
Bristol BS8 1TL, United Kingdom

ABSTRACT

An excellent material for thermal neutron detectors is α-rhombohedral boron, due to the large neutron capture cross section of [10]B, high hole mobility and ability to self-heal from radiation damage, to date, little work has been done on the crystal growth of α-rhombohedral boron. In this investigation, we attempt to grow α-rhombohedral boron by the solution growth method, employing copper as solvent. Well-faceted transparent red crystals several hundreds of microns in size have been made. Elemental analysis of the crystals detected boron, with negligible amounts of copper, suggesting that copper is a promising solvent for the crystal growth of α-rhombohedral boron crystal.

INTRODUCTION

An excellent candidate for thermal neutron detectors is α-rhombohedral boron. The boron 10 isotope has a much larger neutron capture cross section (3840 barns for a 0.025eV thermal neutrons) compared to most other elements (typically ≤ 1 barn) [1] . B^{10} undergoes two nuclear reactions with thermal (~25 meV) neutrons:

(1) $^{10}B + n \rightarrow\ ^7Li$ (0.84 MeV) + ^4He(1.47 MeV) + γ(0.48 MeV)   (94%)

(2) $^{10}B + n \rightarrow\ ^7Li$ (1.02 MeV) + ^4He(1.78 MeV)   (6%)

In both reactions, high energy Li and He (alpha particles) ions are produced. If this process occurs in a semiconductor, for each neutron captured, a huge number (~1.5×10^6) of electron-hole pairs are produced as the energetic ^7Li and ^4He ions created during reactions (1) and (2) pass through the material[2,3] . This charge is sufficiently large to be detected directly without further amplification. Thus, a thermal neutron detector could be based on a boron-rich semiconductor Schottky, *pn*, or *pin* diodes.

Another advantageous property of α-rhombohedral boron is the ability to self heal from neutron bombardment. Boron forms chemical bonds that are different from those common to most compounds, ie the sharing electrons between two atoms. Boron form bonds in which electrons are equally shared between three boron atoms. Employing the three-atom bonds, boron can form icosahedra, twelve atom clusters with boron atoms positioned at each vertex. These boron icosahedra are a basic building block for the crystal structures of pure boron [4]. Thus, α-B semiconductors are suitable for devices, e.g. neutron detectors, subjected to intense radiation since its electrical properties would be unaffected.

Compared with the common β-rhombohedral boron, the rare α-rhombohedral boron is more useful as a semiconductor. The reported hole mobility of α-rhombohedral boron is high, 100 cm^2/V·s, due to its simple crystal structure (12 atoms per unit cell) [5]. In contrast, the hole mobility of β-rhombohedral boron, is two orders of magnitude lower, because of its complex crystal structure (105 atoms per unit cell) which leads to high concentration of inherent structural disorder (a high density of boron interstitials and vacancies). Consequently, the electrical properties of β-B are similar to amorphous semiconductors even though it is a crystal.

α-B is only stable below 1100 °C; at higher temperatures β-B is the stable structure. The upper temperature limit hampers the growth of α-B crystals because of slow kinetics and diffusion rates. There is little work on crystal growth of α-rhombohedral boron. Platinum was a successful solvent for solution growth of α-rhombohedral boron, but the crystal size was small in the previous study (the largest being tenths of millimeters.)[6]. F. Wald and J. Bullitt[7] grew α-rhombohedral boron crystal by chemical vapor deposition. Boron tribromide was chosen as the boron source, tantalum is used as substrate materials. The principal difficulty of the direct CVD growth of α-rhombohedral-boron is that the nucleation is difficult to control: the temperature is much below 1170 °C, crystals of red boron are not formed at all. If the temperature is much hotter, many small crystals occur. The largest crystal of red boron was about 0.1mm. In an effort to promote nucleation at lower temperatures, a modification of CVD growth including a vapor- liquid- solid (VLS) mechanism, was also used by Wald and Bullitt [7]: boron tribromide was reduced by hydrogen on the surface of a liquid solvent, such as Pt, Ag, Au, or Cu. The boron dissolves in and saturates the liquid phase, and then precipitates. The limitation of VLS method is that the nucleation cannot be controlled, even at the lowest temperatures, so that only very small crystals and whiskers grow. Wald and Bullit [7] also attempted to grow α-rhombohedral boron by THM (traveling heater method); unfortunately, α-rhombohedral boron was not successfully produced with this method.

EXPERIMENTAL DETAILS

In this study, the growth of large boron crystals by "solution growth method" was attempted. In this method, a single phase liquid is formed containing the solvent (molten metal) and the solute, boron. Supersaturation of the solute was achieved by cooling the system slowly. Crystals randomly nucleated, then grew by diffusion of the solute to the crystal interface, There are several advantages in solution growth methods. (1) It is a relatively low temperature process, which depends on the solubility of a given material in a solvent, not on the melting point of the material itself. Thus, the goal was to prevent the formation of β-B by using the solvent which can dissolve boron at the temperature lower than 1100 °C. (2)The successful growth of crystals is not as dependent on variables such as minor temperature transients and vibration as other forms of growth which depend on direct solidification from the melt. (3)It is a simple method to operate and, as a result, requires a minimum of operator attention during the experiment.

According to the phase diagram [8] of the boron-copper system, copper is a suitable solvent for the solution growth of boron crystal. First, the eutectic temperature is relatively low, 996 °C, lower than the maximum temperature (about 1100 °C) at which α-B is thermodynamically stable. In fact, F. Wald's reported that the eutectic in the copper boron system showed red boron, and an alloy with 90 at %B clearly showed red boron inclusions in an entirely unchanged microstructure after annealing for 21 days at 1000 °C [9]. Second, there are no intermediate phase boride-copper compounds. Besides, copper is easy to be removed from crystals with concentrated nitric acid. A complete description of the growth of α-boron crystal from copper-boron system is given below.

Solutions were made in a high temperature tube-furnace provided with molybdenum disilicide heating elements, which is capable of reaching temperatures up to 1500 °C in air. The reactor tube was Al_2O_3, and the sample holder is made of pNB coated graphite. In a typical experiment, a mixture of 1g boron (purity 99.5%) and 25g copper (purity 99.99%) was heated at 1350^0C for 5 hours in an argon atmosphere to dissolve the boron into copper. The temperature was reduced from 1350^0C to 1100 °C in half an hour. Then the temperature was slowly reduced from 1100 °C to 996^0C at the cooling rate of 0.5^0C/hour. The boron crystals were grown in this temperature range. At last, the temperature is reduced from 996^0C to room temperature as quickly as possible. The crystals were isolated by etching away the copper with nitric acid

Optical microscopy and SEM (Scanning Electron Microscopy) were employed for observing the surface morphology of boron crystals.

Elemental analysis was done using energy dispersive x-ray spectroscopy (EDS). Elements present in the material are identified by analyzing the energy of the x-rays emitted in response to being bombarded with electrons.

We applied Raman Spectroscopy to determine if α-rhombohedral boron was present. Raman spectra were recorded using a Renishaw inVia spectrometer with the 488 nm Ar^+ laser line as the excitation source. All the measurements were performed in backscattering geometry with unpolarized detection at room temperature.

DISCUSSION

A series of crystals were grown over a range of initial boron concentrations (4 to 20 mole %) and cooling rates. Well-faceted boron crystals in shapes of hexagons (Fig. 1 and 2), rhombus (Fig. 3 and 4), pentagons (Fig. 5 and 6) and triangles and in sizes as large as 500 microns were produced. Some of the crystals were red, suggesting the formation of α-rhombohedral boron (β-rhombohedral boron is black). At relatively low cooling rate (0.5 °C/hour), most of the crystals were in the shape of hexagon or pentagon, at higher cooling rate (1°C/hour), crystals were mostly in the shape of rhombus or triangle. And generally speaking, at slower cooling rates, the crystals were bigger and better faceted.

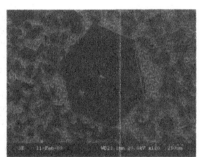

Figure 1. Optical micrograph of hexagonal red boron crystal embedded in copper.

Figure 2. SEM image of the same boron crystal shown in fig.1. The maximum dimension is 500 μm.

Figure 3. Optical micrograph of red Rhombohedral crystal precipitated from the copper-boron melt .

Figure 4. SEM image of the same crystal as shown in fig.3.the largest length of the rhombus is about 150μm.

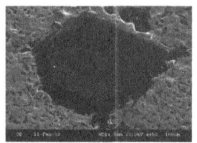

Figure 5. Optical micrograph of red pentagon crystal precipitated from the copper-boron melt.

Figure 6. SEM image of the same crystals as shown in fig. 5. The largest length of the pentagon is about 180μm

The EDS spectra reveals that the samples are 97 wt% boron and 3 wt% copper (99 atomic%, and 1 atomic% respectively). However, the crystals were still incorporated into the copper ingot when the analysis was taken; which could account for some of the copper contamination because the electron beam could have been penetrating the crystal into the copper matrix.

Fig. 7 shows a typical Raman spectrum taken from the boron crystals grown from copper solvent. The Raman spectrum exhibits a narrow band at 528 cm^{-1}, which has been reported for alpha-rhombohedral boron (α-B) crystals [10-12]. This mode has been attributed to the rigid rotation of boron icosahedra and has E_g symmetry [12]. Broad features at 694 and 930 cm^{-1} are related to the vibrational modes of the icosahedra and have A_g symmetry [11, 12]. Additional features with frequencies at ~375, ~820, ~973, and ~1065 cm^{-1}, which have not been reported for α-B, are also observed. A comparison with the previous studies by Richter *et al.* [13] on beta-rhombohedral boron (β-B) shows that these modes are likely from β-B. This indicates that our boron crystals grown from copper solvent consist of both α-B and β-B. A more detailed study has to be carried out in order to fully understand the growth mechanism and optimize the growth conditions of alpha-rhombohedral boron.

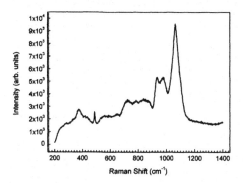

Figure 7. Raman spectra of boron crystal grown from a boron-copper solution.

CONCLUSIONS

Boron crystals were successfully grown using copper as the solvent. By optimizing the parameters important in solution growth, including highest temperature, temperature profile, cooling rate, it should be feasible to grow macrocrystalline or even single crystalline alpha boron.

ACKNOWLEDGEMENTS

The support for this work from the II-VI Inc. Foundation is greatly appreciated.

REFERENCES

1. G.F.Knoll, *Radiation Detection and Measurements* 3rd ed (j. Wiley 2000) p.p 505-535
2. Z.W. Bell, D.A. Carpenter, S.S. Cristy, V.E. Lamberti, A. Burger, B.F. Woodfield, T. Niedermayr, I.D. Hau, S.E. Labov, S. Friedrich, W.G. West, K.R. Pohl, and L. van den Berg, *Phys. Stat. Sol.* C 2 1592 (2005).
3. C. Lundstedt, A. Harken, E. Day, B.W. Robertson, and S. Adenwalla, Nucl. Inst. Meth. A **562** 380 (2006).
4. M. Carrard, D. Emin, L. Zuppiroli, *Phys. Rev.* B **51** 11,270 (1995).
5. O.A Golikova *Chemtronics* **5** ,3(1991)
6. F. Hubbard Horn, *J. Electrochem. Soc.,* Volume 106, Issue **10**, pp. 905-906 (1959)
7. F. Wald, J.Bullitt, "Semiconductor Neutron Detectors". Defense Technical Information Center, Jan 1973, Accession Number: AD0771526.
8. K. T. Jacob, S. Priya, and Y. Waseda, *Metall. Mater. Trans.* A **31**, 2674 (2000).
9. F. Wald, *Electron Technology, Institute of Electron Technology P.A Sci., Warsaw* 3, **1/2**,103 (1971)
10. D.R. Tallant, T.L. Aselage, A.N. Campbell, and D. Emin, *Phys. Rev.* B **40**, 5649 (1989).
11. C.L. Beckel, M. Yousaf, M.Z. Fuka, S.Y. Raja, and N. Lu, *Phys. Rev.* B **44**, 2535 (1991).
12. N. Vast, S. Baroni, G. Zerah, J.M. Besson, A. Polian, M Grimsditch, and I.C. Chervin, *Phys. Rev. Lett.* **78**, 693 (1997).
13. W. Richter, A. Hausen, and H. Binnenbruch, *Phys. Stat. Sol.* (b) **60**, 461 (1973).

Scintillator I

Mater. Res. Soc. Symp. Proc. Vol. 1164 © 2009 Materials Research Society

Scintillating Metal Organic Frameworks: A New Class of Radiation Detection Materials

M. D. Allendorf, R. J. T. Houk, R. Bhakta, I. M. B. Nielsen, F. P. Doty
Sandia National Laboratories
Livermore, CA 94551-0969

ABSTRACT

The detection and identification of subatomic particles is an important scientific problem with implications for medical devices, radiography, biochemical analysis, particle physics, and astrophysics. In addition, the development of efficient detectors of neutrons generated by fissile material is a pressing need for nuclear nonproliferation and counterterrorism efforts. A critical objective in the field of radiation detection is to develop the physical insight necessary to rationally design new scintillation materials for specific applications. However, none of the material types currently used in has sufficient synthetic versatility to exert systematic control over the factors controlling the light output and its dynamics. Here we describe a spectroscopic investigation of two stilbene-based metal-organic frameworks (MOFs) we synthesized, demonstrating that they emit light in response to ionizing radiation, creating the first completely new class of scintillation materials since the advent of plastic scintillators in 1950. This highly novel and unexpected property of MOFs opens a new route to rational design of radiation detection materials, since the spectroscopy shows that both the luminescence spectrum and its timing can be varied by altering the local environment of the chromophore within the MOF. Therefore, the inherent synthetic flexibility of MOFs, which enables both the chromophore structure and its local environment to be systematically varied, suggests that this class of materials can serve as a controlled "nanolaboratory" for probing a broad range of photophysical and radiation detection phenomena. In this presentation we report on the time-dependent fluorescence and radioluminescence of these MOFs and related structures. Multiple decay characteristics have been observed for some materials under study, including fast (ns) exponential and slow (microsecond) non-exponential components. We interpret the results in terms of the electronic states, crystal structures, intermolecular interactions, and transport effects mediating the luminescence. The potential for particle discrimination schemes and large scale production of MOFs and will be discussed.

INTRODUCTION

Conventional organic scintillators (liquid, plastic, or crystalline) suffer from low and highly nonlinear light yield for recoil protons below 10 MeV, resulting in low signals relative to electrons. This effect contributes to a tradeoff between sensitivity and an effective threshold for gamma rejection near the fission peak [1,2]. Pulse shape discrimination (PSD) is characterized by a figure of merit M, and as demonstrated by Moszynski et al. M degrades rapidly with increasing volume, which is a shortcoming of PSD materials and techniques.[3] In liquid scintillators (NE213, EJ301), the neutron path length for elastic scattering on hydrogen at the fission spectrum mean energy of 1.52 MeV is about 10 cm. The typical 5 cm practical depth for liquid scintillators is considerably less than this length; thus, the ultimate sensitivity of liquid scintillators is limited by both the effective threshold and the practical volume for PSD.

Organic scintillators have low luminosity compared with inorganic crystals [1]. This luminosity is further reduced by dE/dx quenching, which causes background Compton events to have an order of magnitude higher light yield than recoil protons at typical PSD thresholds. The brightest organic scintillators, anthracene and stilbene, are not used commercially because of their poor mechanical properties and the difficulty of creating large crystals of these anisotropic materials. It is thus clear that new scintillator materials exhibiting higher and more linear luminosity and/or improved signal characteristics for more effective particle discrimination are needed.

Recently we reported that certain members of a novel class of coordination polymers known as metal-organic frameworks (MOFs) scintillate in response to ionizing radiation [4]. MOFs are a diverse class of crystalline solids, incorporating both inorganic and organic components, and in many cases nanoporosity. Since many factors affect the quantum efficiency and timing of scintillator light output, including chemical composition, electronic structure, inter-chromophore interactions, crystal symmetry, and atomic density, this synthetic flexibility presents an opportunity to exert systematic control of crystal structure, fluor chemistry, and density to an extent unmatched by any other class of material. As a result, MOFs represent an entirely new class of solid scintillators with the potential for neutron detection capabilities exceeding those of existing scintillators.

EXPERIMENTAL METHODS

Crystals of the 2D and 3D MOFs were synthesized using published procedures.[5] Single crystals of each material were mounted on Mylar or gold foils, or on graphite meshes, by placing a drop of a chloroform solution containing crystals and allowing the solvent to evaporate under N_2 atmosphere. In addition, crystals of the stilbenedicarboxylic acid linker (Alfa Aesar) and of anthracene (Aldrich, scintillation grade) were mounted in a similar fashion.

Procedures for the ion-beam induced luminescence (IBIL) measurements may be found in Ref. [4]. The configuration used to detect the scintillation output of the MOFs in response to alpha-particle radiation is shown in Fig. 1. An [241]Am source (3389 Bq; 5 MeV particle energy; Eckert & Ziegler Isotope Products, Valencia, CA) was located above a distribution of MOF particles on a glass slide. A gold-leaf foil was inserted between the source and the particles to reflect upwardly scattered light onto the photomultiplier (Acton Research photon counting module, model PD473), located below the sample. The sample and photomultiplier were enclosed within a housing purged with dry nitrogen to prevent degradation of the 3D MOF by atmospheric water. Individual pulse data were stored and analyzed using a digital oscilloscope (Lecroy Waverunner 6500, 500 MHz), and pulse height spectra were acquired with an Ortec multichannel analyzer and software. Scintillation light was collected for 65 minutes using an Ortec 142A charge-sensitive preamplifier and a shaping amplifier (Canbera 2020) with the shaping time set to 2 μs. Background spectra acquired for equal time with the source over an empty cell were subtracted from the spectra acquired from the MOF samples.

The time dependence of scintillation light was measured using the configuration shown in Fig. 2. The timing resolution for this experiment was determined by the particle detector used. A Hamamatsu 1224 PIN diode with system resolution <200 ps was used in alpha particle experiments. For efficient low-energy beta counting a CANBERRA passivated implanted planar silicon (PIPS) detector was used. A ^{137}Cs beta source was used.

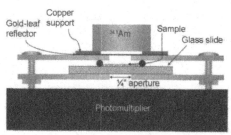

Figure 1. Schematic of experiment used to measure scintillation light output.

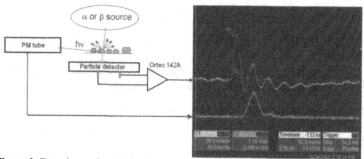

Figure 2. Experimental configuration used to determine the time dependence of MOF scintillation light.

THEORETICAL METHODS

Excitation and emission spectra were computed for the trans-stilbene dicarboxylate linker and for two functionalized linkers. The functionalized linkers incorporate two nitro and amine groups, respectively, as shown in Figure 3. For comparison, spectra were computed for trans-stilbene as well. Optimum geometries were determined for the ground electronic state and the first excited singlet state for all species using the Hartree-Fock method [5] and the configuration interaction singles (CIS) method [6], respectively. In both cases the 6-31G* basis [7] was employed. Vertical excitation end emission energies were computed at these geometries using B3LYP-based time-dependent density functional theory (TDDFT(B3LYP)) with the 6-31+G* basis [8]; additionally, using the same geometries, vertical transition energies were determined with the TDDFT(B3LYP) [9-11] and CIS methods employing the 6-31G* basis and with the semi-empirical ZINDO method [12]. All computations were performed with the Gaussian03 program package [13].

83

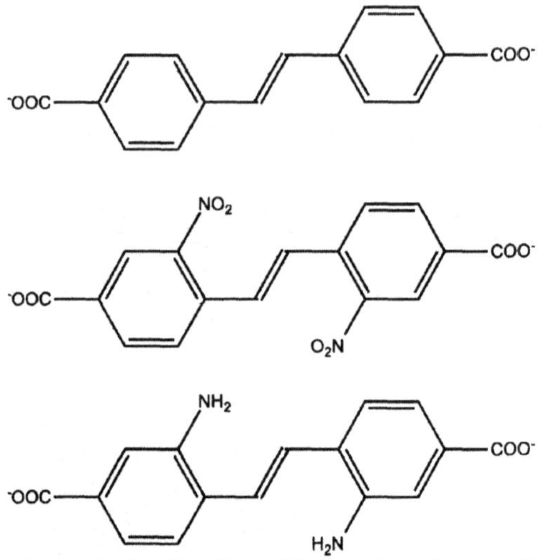

Figure 3. The trans-stilbene dicarboxylate linker (SDC) and two functionalized linkers for which excitation and emission spectra were computed.

METAL-ORGANIC FRAMEWORKS AS SCINTILLATOR MATERIALS

Introduction to metal-organic frameworks

MOFs were first described in the mid 1990s[14,15] and only very recently have their unique properties for radiation detection become apparent. MOFs are a diverse class of hybrid inorganic-organic crystalline materials with chemical functionalities and structures that enable them to be tailored for a wide range of nuclear detection applications. They couple inorganic clusters with organic "linker" groups to create rigid structures that can contain pores or channels that do not collapse upon removal of guest molecules [16-18]. Nanoporous MOFs have monolithic pore dimensions, and unlike zeolites can be prepared in bulk without collapsed "dead" regions, leading to the highest surface areas ever recorded [19]. Some of the most exciting compounds for nuclear detection applications are the "isoreticular MOFs" (IRMOF), in which tetrahedral Zn_4O clusters at the vertices of cubic lattices (Figure 4) are coordinated by a wide range of carboxylate linkers [16,20]. However, MOFs incorporating many other metals have been synthesized, including Mn, Co, Ni, Fe, Mo, Cu, Ag, Cd, In, Sn, lanthanides, and U [18]. IRMOFs with functionalized linkers capable of stabilizing adsorbates, including aromatic rings, NH_2, alkoxy groups, and halogens, can be generated using flexible and straightforward solution-phase methods [21]. Post-synthetic modification is also feasible [22-26]. Thus, facile manipulation of pore size and coordination sphere is feasible.

The permanent nanoporosity in MOFs results in exceptionally high surface areas in some cases (5,900 m^2/g, the highest of any crystalline material,[19] creating the ability to adsorb gases and large organic molecules such as C_{60} and various organic dyes.[27] Exchange of guest

84

molecules is reversible [28-30], supporting the concept of using MOFs as a stabilizing lattice for molecules to promote energy transfer and promote particle capture, or act as wavelength shifters.

Our group is developing MOFs for a variety of sensing applications. This work includes synthesis[28], steady-state and time-dependent spectroscopy [4,28], mechanical properties [31], modeling[31-33], and demonstration of sensing concepts [34]. We also recently published a review of luminescent MOFs [35]. Our recent developments, in addition to MOF-based radiation detection materials, include the first integration of an MOF with a device of any kind to create a functioning sensor [34] and the first "flexible force field" that simulates structural changes induced by gas adsorption in MOFs [32].

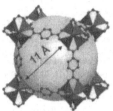

Figure 4. Structure of one pore of the MOF IRMOF-1. Blue tetrahedra: Zn(II) ions; Red: oxygen atoms; black: carbon atoms. Yellow sphere indicates Van der Waals space accessible to guest molecules. Adapted from Ref. [36].

Fluorescent MOFs as scintillator materials

We recently discovered that two new fluorescent MOFs we prepared scintillate in response to ionizing radiation [4]. As part of a program to develop MOFs for chemical sensing, two new fluorescent MOFs incorporating stilbene-dicarboxylate (SDC) linker groups were created [28]. One of these MOFs (MOF-S1) has a nanoporous, three-dimensional framework analogous to Fig. 4 and has a 580 m^2/g surface area. This material has a cubic crystal structure, making it many ways a solid, cubic form of stilbene. A two-dimensional, fully dense analogue was also synthesized (MOF-S2). Both fluoresce when illuminated with ultraviolet (UV) light (Figure 5). In addition, when exposed to the 3 MeV proton beam of an ion microprobe, these MOFs produce ion-beam-induced luminescence (IBIL; Figure 5, dashed curves). They also scintillate when excited by beta and alpha particles. The IBIL intensity is comparable to some commercial organic scintillators (e.g., 22% of anthracene for MOF-S2).[37,38] Preliminary spectroscopic investigations demonstrate the unusual property of particle-dependent luminescence (i.e., the photon-induced luminescence and the IBIL do not overlap; Figure 5), and the extent of this overlap depends on the crystalline environment of the linker. This suggests that MOF scintillator emission wavelengths could be adjusted. Our hypothesis is supported by the results of our related research concerning luminescent MOFs, demonstrating that both the lifetime and spectrum can be rationally manipulated by synthetic means [28,35].

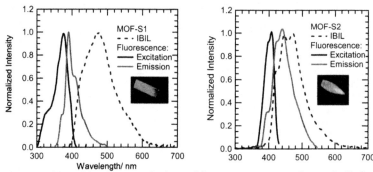

Figure 5. Comparison of the photon excitation and fluorescence spectra of two scintillating MOFs with the corresponding IBIL spectrum [4]. Inset shows single-crystal fluorescence generated by 325 nm excitation[28].

These materials are highly resistant to radiation damage, as shown in Figure 6. Radiation does up to 400 Mrad result in a decrease in the IBIL signals of the two MOFs of about 68%. These doses are many orders of magnitude higher than an anticipated lifetime in-service dose. In contrast, the signal of an anthracene standard lost more than 90% of its initial intensity after on a 50 Mrad dose. The low density of MOF-S1 (0.50 g/cm³) should enhance its resistance to damage compared with anthracene (1.28 g/cm³), since the energy loss per unit length of a material dE/dx scale with its density. Since MOF-S2 has an even higher density (1.52 g/cm³), it is clear that other factors are at work. We suspect that by pinning the carboxylate ligands to the zinc metal centers avoids the formation of photoproducts and cycloadditions, both which are known to decrease the brightness and stability of stilbene.

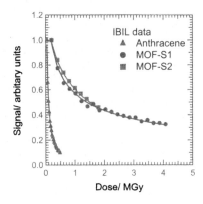

Figure 6. Decay of IBIL signal as a function of dose, comparing the stilbene MOF compounds with anthracene. Solid lines are fits of the maximum IBIL intensity (anthracene: 421 nm; MOF-S2: 473 nm; MOF-S1: 479 nm) to a stretched exponential function.

86

Timing of MOF scintillation light output

Our early scintillation testing with alpha particles shows that both prompt and delayed emission are produced by MOF-S2 (Figure 7), indicating that PSD could be feasible with MOFs [4]. Therefore, the decay characteristics were evaluated for MOF-S2 samples excited by isotopic alpha and beta sources, as shown in Figure 8. The timing resolution for this experiment was determined by the particle detector used. A Hamamatsu 1224 PIN diode alpha detector with system resolution <200 ps shows the dominant fast exponential decay times < 10 ns, followed by a significant non-exponential tail. For efficient low-energy beta counting a CANBERRA passivated implanted planar silicon (PIPS) detector was used. Both curves show the fast exponential components followed by power law decay with significant intensity to > 100 ns. As seen by the fits, the relative intensities of the delayed component are 32% for alphas and 12% for betas, which can be compared directly with published values of 20% and 3.5% for anthracene [39], proving that PSD should be feasible with this MOF. We note that experiments with MOF-S1 (data not shown) show that this material displays a rather different time signature. The exact origins of this difference are not completely clear. However, our comparison of IBIL data with the fluorescence spectra of these two MOFs strongly suggests that this is related to the extent of interchromophore overlap[4]. Since this can be controlled to some degree by altering geometric and electronic structure of the organic linker group in the MOF, it should be possible to determine a set of rational design principles for tuning PSD within MOFs.

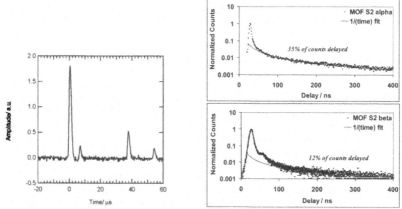

Figure 7 (left). Oscilloscope trace showing prompt and delayed emission from MOF-S2 in response to alpha particles. **Figure 8 (right).** PSD exhibited by MOF-S2, showing clearly different time signatures for alpha and beta particles.

QUANTUM-CHEMISTRY MODELING OF MOF ELECTRONIC STRUCTURE

Complementing and supporting our synthetic and spectroscopic investigations is a parallel theoretical effort to enable interpretation of the experimental results and provide the basis for the eventual development of simulation tools capable of designing new MOF-based scintillation materials. We are currently computing the ground- and lowest-excited singlet and triplet potential energy surfaces for MOF linkers incorporating stilbene and anthracene groups using quantum chemical (QC) methods. Methods that are both applicable to large molecular systems and capable of handling excited states, such as configuration interaction with single excitations (CIS) and time-dependent density functional theory (TDDFT) are appropriate for this purpose. Both the CIS and TDDFT methods have been applied extensively for investigations of excited electronic states, including the computation of luminescence spectra for aromatic hydrocarbons [40-42]. We are computing equilibrium geometries, harmonic vibrational frequencies, and, to obtain spectral intensities, Franck-Condon (FC) overlap factors and oscillator strengths. To our knowledge this is the first application of QC methods to predict MOF fluorescence intensities and will provide a challenging test of theory. Because of the extended structure of MOFs, Zn-linker clusters that include the essential symmetry elements will be used as MOF analogues. Since even these can be large, we are evaluating the performance of less expensive semi-empirical methods such as INDO or CNDO.

Preliminary high-level QC calculations indicate that adding substituent's to the MOF linker groups should have a substantial effect on the excitation and emission wavelengths. In Figure 9 predicted excitation and emission energies are shown for trans-stilbene dicarboxylate (SDC) plots) and for SDC functionalized with amino and nitro groups. These energies represent vertical transitions between the ground vibronic state of the molecule and its first excited singlet state. They were computed at the TDDFT(B3LYP)/6-31+G* level of theory, which is the highest level of theory we have used to date. We therefore expect these results to be quite accurate, based on comparison with the known spectra of trans stilbene and SDC. The vertical transition energies for trans-stilbene and SDC are similar, indicating that the carboxylate groups in SDC do not significantly perturb the pi-electron system on the stilbene unit; this is also borne out by the very similar geometries found for the carbon framework in the two systems. The effect on the transition energies of adding amino groups appears to be correlated with the corresponding changes in the geometry for the ground and excited electronic states. A (nearly) planar geometry is found for both the ground and excited electronic state of SDC as well as for the ground state of the amino-substituted SDC, but the latter has a nonplanar excited state in which the angle between the planes of two phenyl rings is around 45 degrees. Accordingly, addition of amino groups is found to have a negligible effect on the excitation energy but to cause a significant lowering of the emission energy.

In contrast, the electron-withdrawing power of the nitro groups results in large shifts of both the excitation and emission energies. Such modifications to the MOF linker groups are synthetically feasible and demonstrate that tuning of the luminescence spectrum within a MOF scintillator should be feasible. We are currently synthesizing these modified linker groups and their associated MOFs.

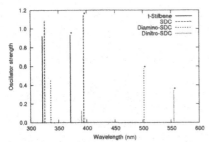 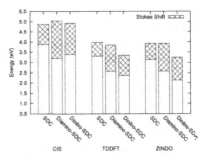

Figure 9 (left). TDDFT(B3LYP)/6-31+G* vertical excitation and emission energies for trans-stilbene, the trans-stilbene dicarboxylate linker (SDC), and two functionalized linkers. Emission energies are labeled with an asterisk.

Figure 10 (right). Comparison of vertical transition energies obtained with the CIS, TDDFT, and ZINDO methods. The top and bottom edges of the cross-hatched part of each column represent the excitation and emission energies, respectively.

Since MOFs themselves are too large to treat by most QC methods, we compared the results of TDDFT, CIS, and semiempirical QC method known as ZINDO (Zerner's Intermediate Neglect of Differential Overlap). Although of lower absolute accuracy, ZINDO can address much larger molecular systems than either of the other two methods, which should allow us to model MOF subunits of at least the size of the so-called "secondary building unit." A comparison of Stokes shifts (the energy difference between the excitation and emission spectra) predicted by the three methods is shown in Figure 10. The important conclusion to be taken from these results is that ZINDO accurately reproduces the trends of the higher-level methods. It should therefore be useful for modeling the effects perturbations to the MOF structure, which will provide useful guidance to our efforts to optimize the light output characteristics of MOF scintillators.

CONCLUSIONS

MOFs represent a completely new class of scintillation materials with unique properties that enable a level of synthetic control over their electronic structure that is unavailable in conventional scintillators. MOFs can therefore serve as a controlled "nanolaboratory" for probing a broad range of photophysical phenomena. In addition, their nanoporosity provides new opportunities to enhance their capabilities by, for example, absorbing compounds that shift emission wavelengths to more convenient regions of the spectrum. From a practical point of view, our investigations show that MOFs can have light outputs comparable to commercial organic scintillators and are also highly damage resistant. Luminescence timing data that we report here demonstrate that PSD is feasible.

This work was supported by the Sandia National Laboratories Laboratory Directed Research and Development program and by the Defense Threat Reduction Agency under contract 074325I-0.

REFERENCES

1. Klein, H.; Brooks, F. D. "Proc. Int. Workshop Fast Neutron Detectors Appl.", 2006, POS(FNDA2006). Available at http://pos.sissa.it.
2. Moszynski, M.; Costa, G. J.; Guillaume, G.; Heusch, B.; Huck, A.; Mouatassim, S. *Nucl. Inst. Meth. Phys. Res. A* **350**, 226 (1994).
3. Knoll, G. F. *Radiation Detection and Measurement*, 3rd ed.; Wiley: New York, 2000.
4. Doty, F. P.; Bauer, C. A.; Skulan, A. J.; Grant, P. G.; Allendof, M. D. *Adv. Mater.* **21**, 95 (2009).
5. Szabo, A.; Ostlund, N. S. *Modern Quantum Chemistry*, 1st revised ed. ed.; McGraw-Hill: New York, 1989.
6. Foresman, J. B.; Head-Gordon, M.; Pople, J. A.; Frisch, M. J. *J. Phys. Chem.* **96**, 135 (1992).
7. Hariharan, P. C.; Pople, J. A. *Theor. Chim. Acta* **28**, 1973 (1973).
8. Ditchfield, R.; Hehre, W. J.; Pople, J. A. *J. Chem. Phys.* **54**, 724 (1971).
9. Bauernschmitt, R.; Ahlrichs, R. *Chem. Phys. Lett.* **256**, 454 (1996).
10. Becke, A. D. *J. Chem. Phys.* **98**, 5648 (1993).
11. Lee, C. T.; Yang, W. T.; Parr, R. G. *Phys. Rev. B* **37**, 785 (1998).
12. Zerner, M. C. *Rev. Comp. Chem.*; VCH: New York, 1991; Vol. 2.
13. M. J. Frisch, G. W. T., H. B. Schlegel, G. E. Scuseria,; M. A. Robb, J. R. C., J. A. Montgomery, Jr., T. Vreven,; K. N. Kudin, J. C. B., J. M. Millam, S. S. Iyengar, J. Tomasi,et al. Gaussian 03, Rev. C.02; Gaussian, Inc.: Wallingford, CT, 2004.
14. Yaghi, O. M.; Li, G. M.; Li, H. L. *Nature* **378**, 703 (1995).
15. Yaghi, O. M.; Li, H. L. *J. Am. Chem. Soc.* **117**, 10401 (1995).
16. Eddaoudi, M.; Moler, D. B.; Li, H.; Chen, B.; Reineke, T. M.; O'Keefe, M.; Yaghi, O. M. *Acc. Chem. Res.* **34**, 319 (2001).
17. Ferey, G. *Chem. Mater.* **13**, 3084 (2001).
18. James, S. L. *Chem. Soc. Rev.* **32**, 276 (2003).
19. Ferey, G.; Mellot-Drazniek, C.; Serre, C.; Millange, F.; Dutour, J.; Surble, S.; Margiolaki, I. *Science* **309**, 2040 (2005).
20. Rowsell, J. L. C.; Yaghi, O. M. *Angew. Chem. Int. Ed.* **44**, 4670 (2005).
21. Eddaoudi, M.; Kim, J.; Rosi, N.; Vodak, D.; Wachter, J.; O'Keefe, M.; Yaghi, O. M. *Science* **295**, 469 (2002).
22. Costa, J. S.; Gamez, P.; Black, C. A.; Roubeau, O.; Teat, S. J.; Reedijk, J. *European Journal of Inorganic Chemistry*, 1551 (2008).
23. Furukawa, S.; Hirai, K.; Nakagawa, K.et al. *Angewandte Chemie-International Edition* **48**, 1766 (2009).
24. Song, Y. F.; Cronin, L. *Angewandte Chemie-International Edition* **47**, 4635 (2008).
25. Tanabe, K. K.; Wang, Z. Q.; Cohen, S. M. *Journal of the American Chemical Society* **130**, 8508 (2008).
26. Wang, Z. Q.; Cohen, S. M. *Journal of the American Chemical Society* **129**, 12368 (2007).
27. Chae, H. K.; Siberio-Perez, D. Y.; Kim, J.; Go, Y.; Eddaoudi, M.; Matzger, A. J.; O'Keeffe, M.; Yaghi, O. M. *Nature* **427**, 523 (2004).
28. Bauer, C. A.; Timofeeva, T. V.; Settersten, T. B.; Patterson, B. D.; Liu, V. H.; Simmons, B. A.; Allendorf, M. D. *J. Am. Chem. Soc.* **129**, 7136 (2007).

29. Cussen, E. J.; Claridge, J. B.; Rosseinsky, M. J.; Kepert, C. J. *J. Am. Chem. Soc.* **124**, 9574 (2002).
30. Yaghi, O. M.; Davis, C. E.; Li, G.; Li, H. *J. Am. Chem. Soc.* **119**, 2861 (1997).
31. Bahr, D. F.; Reid, J. A.; Mook, W. M.et al. *Phys. Rev. B* **76** (2007).
32. Greathouse, J. A.; Allendorf, M. D. *J. Am. Chem. Soc.* **128**, 10678 (2006).
33. Greathouse, J. A.; Allendorf, M. D., Force Field Validation for Molecular Dynamics Simulations of IRMOF-1 and Other Isoreticular Zinc Carboxylate Coordination Polymers ," *J. Phys. Chem. C*, accepted for publication, 2008.
34. Allendorf, M. D.; Houk, R. J. T.; Andruszkiewicz, L.; Talin, A. A.; Pikarsky, J.; Choudhury, A.; Gall, K. A.; Hesketh, P. J. *J. Amer. Chem. Soc.* **130**, 14404 (2008).
35. Allendorf, M. D.; Bauer, C. A.; Bhakta, R. K.; Houk, R. J. T. *Chem. Soc. Rev.* **DOI: 10.1039/b802352m** (2009).
36. Li, H.; Eddaoudi, M.; O'Keeffe, M.; Yaghi, O. M. *Nature* **402**, 276 (1999).
37. Doty, F. P.; Allendorf, M. D., patent pending.
38. Doty, P.; Bauer, C.; Grant, P.; Skulan, A. J.; Allendorf, M. D., unpublished data.
39. Gibbons, P. E.; Northrop, D. C.; Simpson, O. *Proc. Phys. Soc.* **79**, 373 (1962).
40. Dierksen, M.; Grimme, S. *J. Chem. Phys.* **120**, 3544 (2004).
41. Hirata, S.; Lee, T. J.; Head-Gordon, M. *J. Chem. Phys.* **111**, 8904 (1999).
42. Jas, G. S.; Kuczera, K. *Chem. Phys. Lett.* **214**, 229 (1997).

Mater. Res. Soc. Symp. Proc. Vol. 1164 © 2009 Materials Research Society   1164-L07-02

Optimization of Ce Content in $Ce_xLa_{1-x}F_3$ Colloidal Nanocrystals for Gamma Radiation Detection

Nathan J. Withers, Krishnaprasad Sankar, John B. Plumley, Brian A. Akins, Tosifa A. Memon, Antonio C. Rivera, Gennady A. Smolyakov, and Marek Osiński

Center for High Technology Materials, University of New Mexico
1313 Goddard SE, Albuquerque, NM 87106-4343
Tel. (505) 272-7812; Fax (505) 272-7801; E-mail: osinski@chtm.unm.edu

ABSTRACT

We report on experimental investigations of $Ce_xLa_{1-x}F_3$ colloidal nanocrystals (CNCs) and their properties in function of Ce content. The CNCs were characterized by TEM, energy-dispersive X-ray spectroscopy (EDS), steady-state UV-VIS optical absorption and photoluminescence (PL) spectroscopy, and by PL lifetime measurements. We also report on observations of scintillation from the cerium-doped lanthanum fluoride CNC material in experiments on radiation detection.

INTRODUCTION

Colloidal nanocrystals (CNCs) are particles with the dimensions on the scale of 5-50 nanometers across, synthesized by chemical processes. At these scales, the crystals exhibit enhanced quantum mechanical effects such as bandgap widening, decrease in carrier lifetime, and the dominance of surface effects. The chemical assembly of these materials is commonly known as "bottom up" nanotechnology, where self-assembly is used to create large quantities of nanoscale materials that would be prohibitively expensive to create using other techniques such as e-beam lithography. CNCs have attracted tremendous interest over the last few years for a wide range of biomedical, biochemical sensing, and optoelectronic applications [1-3]. For example, this technology is being researched to produce phosphors for light emitting diodes [3-4], markers for biological labeling [5], and photovoltaic devices [6].

Detection of nuclear radiation by its conversion to UV or visible light can be another attractive application of CNCs. The luminescence of lanthanide ions in organic media has been shown to greatly improve by doping them in the inorganic core of nanoparticles that are soluble in organic solvents [7]. This is due to the fact that the long-lived excited state of the lanthanide ions is quenched very effectively by the high-energy vibrations of closely spaced organic groups [8]. Since higher Ce content is expected to lead to increased self-absorption of Ce emission [9], an optimal concentration should exist that maximizes quantum efficiency at a fixed CNC concentration. Due to their small size, CNCs are expected to have better solubility in organic polymer or inorganic sol-gel host materials and to cause much less scattering, which should result in higher efficiency of the scintillator.

There have been very few published preliminary studies of radiation response of nanocomposites based on CNCs. Lithiated (^{6}LiOH) sol-gel doped with CdSe/ZnS QDs was considered as a potential neutron detector and tested under alpha irradiation [10]. Commercial CdSe/ZnS core-shell QDs were inserted in porous glass and exposed to alpha [11] and gamma [12] irradiation. Quantum dot/organic semiconductor composite based on CdSe/ZnS core-shell QDs was tested under electron-beam excitation using cathodoluminescence [13]. LaF_3:Ce CNCs embedded in an organic matrix materials were exposed to gamma irradiation [14]. Very low scintillation output was observed in all the cases, which was ascribed to poor stopping power of unoptimized nanocomposites and low concentration of nanocrystals in liquid solutions.

In the characterization of scintillation materials, there are several criteria used to determine how effective a scintillator is. Common criteria are the efficiency of converting a high-energy photon to UV/Visible light photons, the linearity of this conversion, lack of self absorption of scintillation light, small decay time of scintillation light, ease of manufacture and other parameters [15].

EXPERIMENT

For synthesis of $Ce_xLa_{1-x}F_3$ CNCs, we have used a modified procedure of Wang *et al.* [16] originally developed to produce chitosan-coated Eu-doped LaF_3 CNCs.

Three stock solutions were used as reagents and prepared before the synthesis. Chitosan solution (1 wt%) was prepared by dissolving 1 g of highly viscous chitosan, purchased from Fluka, in a mixture of 0.428 g of 37% HCl solution and 100 ml of deionized (DI) water. Lanthanum chloride stock solution (0.2 M) was prepared by dissolving 17.668 g of lanthanum (III) chloride heptahydrate ($LaCl_3•7H_2O$), purchased from Sigma-Aldrich, in 250 ml of DI water. Cerium chloride stock solution (0.2 M) was prepared by dissolving 3.758 g of cerium (III) chloride heptahydrate ($CeCl_3•7H_2O$), purchased from Sigma-Aldrich, in 50 ml of DI water.

A 500-ml three-neck flask used for the synthesis was kept under argon atmosphere in a Schlenk line setup. The flask was placed on a heating and stirring mantle, controlled by a programmable ramping temperature controller through a type-J thermocouple and a stirring controller. For the synthesis of LaF_3 doped with 5% Ce, 9.5 ml of 0.2 M $LaCl_3$ solution and 0.5 ml of 0.2 M $CeCl_3$ solution were added to 25 ml of chitosan solution and mixed thoroughly. Then, 0.2222 g of ammonium fluoride (NH_4F) in 10 ml of DI water was added to the mixture. The pH of the mixture in the flask was adjusted to 6.5 with diluted ammonia. Subsequently, the flask was purged with argon and closed. The temperature controller was programmed to ramp the temperature of the solution in the flask to 75 °C and maintain that temperature for 2 hours, with constant stirring under argon atmosphere. After cooling to room temperature, the CNCs obtained were precipitated with acetic acid and collected by centrifugation at 4000 rpm. After the supernatant was discarded, the CNCs were re-dissolved in DI water.

$Ce_xLa_{1-x}F_3$ CNCs with various Ce content were synthesized as described above by varying the amount of $CeCl_3$ solution used in the synthesis. While maintaining all other synthesis parameters unchanged, eight different syntheses were performed with the intended Ce content (x) of 0.5%, 5%, 10%, 15%, 20%, 25%, 75%, and 100% with respect to the total lanthanide (cerium + lanthanum) content.

Transmission electron microscopy (TEM) and energy dispersive spectroscopy (EDS) were performed on a high-resolution transmission electron microscope JEOL-2010F operating at 200 kV and equipped with OXFORD Link ISIS EDS apparatus.

Absorption measurements were performed on a Cary 400 UV-VIS spectrophotometer. Photoluminescence (PL) measurements were performed on a Horiba Jobin Yvon Fluorolog-3 spectrofluorometer. PL lifetime measurements were taken on the same spectrofluorometer in a different configuration allowing for time correlated single photon counting. A pulsed LED emitting at 250 nm was the excitation source used for these experiments.

Scintillation measurements were taken with a custom system using a Hamamatsu R7449 quartz-window bi-alkali photomultiplier tube (PMT) at a bias of 910 V. The electronic signal from the PMT was processed using Ortec 113 preamplifier, Ortec 570 amplifier and pulse shaper, and Ortec Illusion 25 multichannel analyzer. Data were analyzed using Ortec Maestro-32 for Windows software. The parameters of the Ortec 570 amplifier were: gain of 200 and a shaping time of 2 μs. All measurements were taken over a live time of 5,000 s. The cuvette that contained the $Ce_xLa_{1-x}F_3$ CNC solution was a Starna Cells 32/Q/20 cylindrical Spectrosil cuvette with an interior diameter of 19 mm and path-length of 20 mm giving a volume of 5.6 mL. All surfaces of the cuvette, except for the front window, were coated with teflon tape to improve light collection. The radioactive source, placed 0.5 inches from the back face of the cuvette, was a bi-energetic Co-57 disk source purchased from Spectrum Technologies with gamma energies of 122.07 keV and 136.48 keV. Both of these energies are too low to cause Cherenkov radiation in water. At the time of measurement, the activity of the source was 0.94 μCi.

DISCUSSION

TEM studies

The TEM image (Fig. 1a) shows the synthesized nanocrystals as hexagonal platelets, 10-12 nm in diameter and 4-6 nm thick, oriented both flat and side-on. The hexagonal structure of the CNCs is in agreement with the reported crystal structure of bulk LaF_3 [17]. The EDS analysis (Fig. 1b) confirms presence of Ce, La, and F in the composition of the synthesized CNCs. The Cu and C peaks originate from the holding grid. The spectrum also shows presence of Cl, which may be due to the residue of some ammonium chloride or lanthanum chloride in the sample.

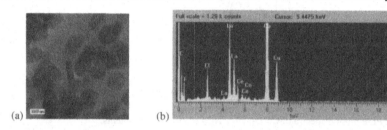

Figure 1. (a) TEM image of $Ce_xLa_{1-x}F_3$ nanocrystals; (b) EDS spectrum of nanocrystals.

Absorption spectroscopy

The absorption measurements show four clearly resolved peaks at 206, 218, 234 and 248 nm as shown in Fig. 2a. These four peaks agree well (Fig. 2b) with the internal $4f \rightarrow 5d$ energy level configuration of Ce^{3+} ions doped into a LaF_3 host matrix [18].

PL spectroscopy and optimization of Ce content

Fig. 3 shows dependence of PL spectra on cerium content. The peak emission intensity and wavelength are plotted for increasing cerium content x in Fig. 4. An optimum is observed at $x =$ 10%. A sharp fall off in intensity is observed for the samples with 5% and 1% cerium concentration, whereas the emission intensity for higher concentration samples decreased only slightly. The peak emission wavelength red shifted with increasing cerium concentration, while an almost constant optimal excitation wavelength of ~250 nm was observed in the PL excitation spectra for all the samples.

95

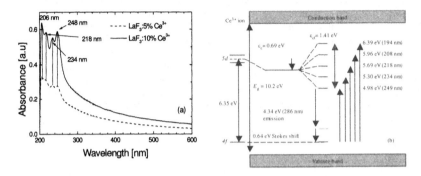

Figure 2. (a) Absorption spectra of $Ce_{0.05}La_{0.95}F_3$ and $Ce_{0.1}La_{0.9}F_3$ CNCs; (b) Internal energy level configuration of cerium ion doped into a lanthanum fluoride host lattice [18].

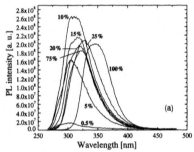

Figure 3. PL emission spectra of $Ce_xLa_{1-x}F_3$ CNCs at various Ce contents.

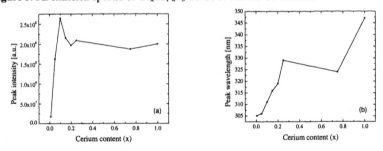

Figure 4. Variation of (a) PL peak intensity and (b) PL peak position with increasing Ce content.

PL lifetime measurements

PL lifetime measurements showed a dominant decay component from 27.3 to 28.8 ns for 5% and 10% Ce content; one of the measurements is illustrated in Fig. 5. This value is in good agreement with the dominant scintillation decay component for bulk $Ce_xLa_{1-x}F_3$ [9]. Although the energy conversion mechanisms are not the same in PL lifetime and scintillation decay measurements, nonetheless the results of the PL lifetime measurements provide a fairly good estimate of how fast $Ce_xLa_{1-x}F_3$ nanoscintillators can be by probing Ce intrinsic lifetime in LaF_3.

Scintillation of $Ce_xLa_{1-x}F_3$

To avoid false positives due to cosmic ray events and scintillation in the quartz, water, and chitosan coating, scintillation tests were carried out separately on 11.6 mg/ml chitosan blanks and the $Ce_xLa_{1-x}F_3$ CNC sample with a loading of 2.7 mg/ml. Although gamma spectra were not resolved, Fig. 6 shows a clear increase in counts for the $Ce_xLa_{1-x}F_3$ CNCs exposed to ^{57}Co.

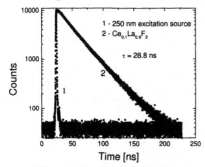

Figure 5. Results of PL lifetime measurement for $Ce_{0.1}La_{0.9}F_3$ CNCs.

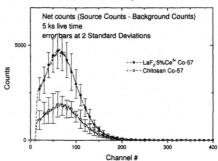

Figure 6. Scintillation of $Ce_{0.5}La_{0.95}F_3$ under Co-57 exposure.

CONCLUSIONS

$Ce_xLa_{1-x}F_3$ CNCs were characterized by TEM, EDS, steady state UV-VIS optical absorption and PL spectroscopy, and by PL lifetime measurements. Experiments on radiation detection were conducted to detect scintillation from the cerium doped lanthanum fluoride CNC material.

In general, our results agree well with previously reported results for bulk Ce-doped LaF_3 scintillators. $Ce_xLa_{1-x}F_3$ CNCs demonstrated high PL output and short PL lifetimes. Based on the results of PL measurements, the optimum 10% Ce content was identified. Scintillation of the $Ce_xLa_{1-x}F_3$ CNC material was confirmed with the use of low-activity monoenergetic γ sources.

ACKNOWLEDGMENTS

This work was supported by the Department of Homeland Security under Grant No. 2008-DN-077-AR1023-02, by the Defense Threat Reduction Agency under Grant No. HDTRA-1-08-1-0021, and by the National Science Foundation under Grant No. IIS-0610201. Nathan Withers is a fellow supported by the National Science Foundation under the IGERT program on Integrating Nanotechnology with Cell Biology and Neuroscience (Grant No. DGE-0549500).

REFERENCES

1. M. Osiński, T. M. Jovin, and K. Yamamoto, (Eds.), *Colloidal Quantum Dots for Biomedical Applications III*, San Jose, CA, 19-21 Jan. 2008, Proc. SPIE **6866**.

2. I. Matsui, "Nanoparticles for electronic device applications: A brief review", J. Chem. Eng. Japan **38** (8), 535-546 (2005).
3. Y.-Q. Li, A. Rizzo, R. Cingolani, and G. Gigli, "White-light-emitting diodes using semiconductor nanocrystals", Microchimica Acta **159** (3-4), 207-215 (July 2007).
4. H. S. Chen and S. J. J. Wang, "White-light emission from organics-capped ZnSe quantum dots and application in white-light-emitting diodes", Appl. Phys. Lett. **86**, Art. 131905 (2005).
5. X. Michalet, F. Pinaud, T. D. Lacoste, M. Dahan, M. P. Bruchez, A. P. Alivisatos, and S. Weiss, "Properties of fluorescent semiconductor nanocrystals and their application to biological labeling", Single Molecules **2** (4), 1438-5171 (2001).
6. M. Shim and P. Guyot-Sionnest, "n-type colloidal semiconductor nanocrystals", Nature **407**, 981-983 (2000).
7. J. W. Stouwdam, G. A. Hebbink, J. Huskens, and F. C. J. M. van Veggel, "Lanthanide-doped nanoparticles with excellent luminescent properties in organic media", Chem. Mater. **15** (24), 4604-4616 (2003).
8. S. I. Klink, G. A. Hebbink, L. Grave, F. G. A. Peters,F. C. J. M. Van Veggel, D. N. Reinhoudt, and J. W. Hofstraat, "Near-infrared and visible luminescence from terphenyl-based lanthanide(III) complexes bearing amido and sulfonamido pendant arms", Eur. J. Org. Chem. **10**,1923-1931 (2000).
9. W. W. Moses and S. E. Derenzo, "The scintillation properties of cerium-doped lanthanum fluoride", Nucl. Instrum. Methods Phys. Res. A **299** (1-3), 51-56 (1990).
10. S. Dai, S. Saengkerdsub, H.-J. Im, A. C. Stephan, and S. M. Mahurin, "Nanocrystal-based scintillators for radiation detection", *Unattended Radiation Sensor Systems for Remote Applications*, 15-17 Apr. 2002, Washington, DC, AIP Conf. Proc. **632**, pp. 220-224, 2002.
11. S. E. Létant and T.-F. Wang, "Study of porous glass doped with quantum dots or laser dyes under alpha irradiation", Appl. Phys. Lett. **88** (10), Art. 103110, 8 March 2006.
12. S. E. Létant and T. F. Wang, "Semiconductor quantum dot scintillation under γ-ray irradiation", Nano Lett. **6** (12), 2877-2880, 13 Dec. 2006.
13. I. H. Campbell and B. K. Crone, "Quantum-dot/organic semiconductor composites for radiation detection", Adv. Mater. **18** (1), 77-79 (2006).
14. E. A. McKigney, R. E. Del Sesto, L. G. Jacobsohn, P. A. Santi, R. E. Muenchausen, K. C. Ott, T. M. McCleskey, B. L. Bennett, J. F. Smith, and D. W. Cooke, "Nanocomposite scintillators for radiation detection and nuclear spectroscopy", Nucl. Instrum. Methods Phys. Res. A **579** (1), 15-18, 21 Aug. 2007.
15. G. F. Knoll, *Radiation Detection and Measurement* (John Wiley and Sons, New York, 2000), p. 219.
16. F. Wang, Y. Zhang, X. Fan, and M. Wang, "One-pot synthesis of chitosan/LaF$_3$:Eu^{3+} nanocrystals for bio-applications", Nanotechnology **17**, 1527-1532 (2006).
17. N. Stubicar, P. Zipper, and B. Cherney, "Variety of aggregation and growth processes of lanthanum fluoride as a function of La/F activity ratio", Crystal Growth & Design **5** (1), 123-128 (2005).
18. P. Dorenbos, "5d-level energies of Ce^{3+} and the crystalline environment. I. Fluoride compounds", Phys. Rev. B **62** (23), 15640-15649 (2000).

Mater. Res. Soc. Symp. Proc. Vol. 1164 © 2009 Materials Research Society   1164-L07-04

Studies of Non-Proportionality in Alkali Halide and Strontium Iodide Scintillators Using SLYNCI

Larry Ahle[1], Gregory Bizarri[2], Lynn Boatner[3], Nerine J. Cherepy[1], Woon-Seng Choong[2], William W. Moses[2], Stephen A. Payne[1], Kanai Shah[4], Steven Sheets[1], and Benjamin, W. Sturm[1]

[1]Lawrence Livermore National Laboratory, 7000 East Avenue, Livermore, CA 94551, U.S.A.

[2]Lawrence Berkeley National Laboratory, 1 Cyclotron Road, Berkeley, CA 94720, U.S.A.

[3]Oak Ridge National Laboratory, Oak Ridge, TN 37831

[4]Radiation Monitoring Devices, Watertown, MA 02472

ABSTRACT

Recently a collaboration of LLNL and LBNL has constructed a second generation Compton coincidence instrument to study the non-proportionality of scintillators [1-3]. This device, known as SLYNCI (Scintillator Light-Yield Non-proportionality Characterization Instrument), has can completely characterize a sample with less than 24 hours of running time. Thus, SLYNCI enables a number of systematic studies of scintillators since many samples can be processed in a reasonable length of time. These studies include differences in non-proportionality between different types of scintillators, different members of the same family of scintillators, and impact of different doping levels. The results of such recent studies are presented here, including a study of various alkali halides, and the impact of europium doping level in strontium iodide. Directions of future work area also discussed.

INTRODUCTION

The achievable energy resolution for a given scintillator material used for gamma ray spectroscopy is principally dependent on the efficiency for conversion of the electrons and holes into scintillation light, or light yield. The efficiency for this process is not independent of energy deposited in the crystal which leads to a non-proportionality between energy deposited in the crystal and the light yield produced. This non-proportionality is often the limiting factor in determining the final resolution of the crystal [4-7]. To measure this proportionality, a collaboration of LLNL and LBNL has constructed a second generation Compton coincidence instrument known as SLYNCI (Scintillator Light-Yield Non-proportionality Characterization Instrument) [1-3]. This instrument has over a 30 times higher data collection rate than previous devices,[7-9] enabling a complete non-proportionality measurement in less than 24 hours. This increase throughput makes it more straight forward to make systematic studies of scintillator and non-proportionality. These studies include differences in non-proportionality between different types of scintillators, different members of the same family of scintillators, and impact of different doping levels. Such studies were performed for several alkali halides, strontium iodide, and lanthanum bromide.

EXPERIMENT

SLYNCI relies on Compton scattering events to measure the non-proportionality of a scintillator. It consists of a collimated ^{137}Cs source, a PMT for light detection from the scintillator being tested, and five HPGe detectors surrounding the scintillator. Gamma rays with energy of 661.660 keV from the ^{137}Cs source are Compton scattered in the scintillator with the scattered photon being detected in one of the HPGe. The energy measured in the HPGe detector then determines how much energy was deposited in the scintillator, while the signal from the PMT measures how much light was produced by the scintillator. With this information for the entire range of Compton electron energies, the non-proportionality curve can be determined.

To cover the entire energy range, two data runs are taken, one with collimator at 0° and one at 15° relative to the center of the middle HPGe detector. The signals from all detectors are fed through charge sensitive preamps and then into 100 MHz waveform digitizers from Struck. A separate FPGA takes the trigger output from the digitizers and processes for coincidence events between the PMT and one of the HPGe. Scaled down singles events from the PMT and the HPGe detectors are also mixed into the data stream for calibration purchases. The PMT and HPGe that sees low angle Compton scattered events are calibrated every roughly every 10 minutes. The other four HPGe are calibrated from a separate calibration run. Signals in valid events are then processed online through a trapezoidal filter to generate signal amplitudes for each triggered detector. Additionally, the raw waveform from the PMT is saved to allow for processing offline.

Alkali Halide Data

Several alkali halides were run through SLYNCI and the data is shown in figure 1. The NaI(Tl) crystal was manufactured by Saint Gobain and is a cylinder with diameter of 12.5 mm and length of 12.5 mm. The CsI(Tl) crystal was also manufactured by Saint Gobain but has a diameter of 20 mm and length of 20 mm. The CsI(Na) crystal was manufactured by ScintiTech and is a cylinder with a diameter of 10 mm and length of 10 mm. All three curves show somewhat similar behavior but note NaI(Tl) has a peak at slightly higher energy and at a lower value, giving NaI(Tl) a slightly better overall non-porportionality.

Strontium Iodide Data

We have begun to study strontium iodide specifically looking at effects due to europium dopant concentration. Figure 2 shows the relative yield for one $SrI_2(Eu)$ crystal grown at ORNL. The dopant concentration is 6% as tests at LLNL have shown this concentration appears to maximize the light yield. The shape of the SrI_2 curve is similar to the alkali halide curves but the amplitude change is significantly less. Specifically, the peak value for the alkali halides is about 15-25% as compared to their value at 446 keV. For the SrI_2 this change is only about 5%. The increased scatter seen in this data is believed to be due to energy escaping from the SrI_2 crystal as this crystal was a cube of roughly 5 mm a side.

100

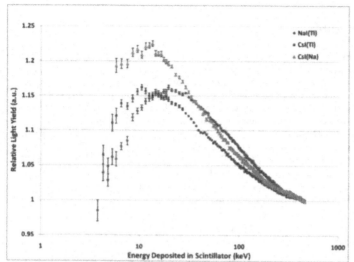

Figure 1. Relative light yield curves for NaI(Tl), CsI(Tl) and CsI(Na). Each curve has been renormalized to its value at 446 keV.

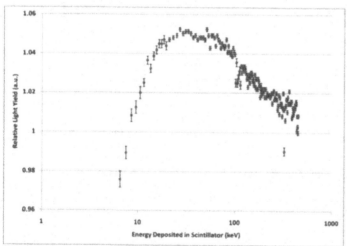

Figure 2. Relative light yield curve for $SrI_2(Eu)$. The dopant level is 6% and the data has been renormalized to its value at 446 keV.

Lanthanum Bromide Data

A study of the impact of cerium dopant level for LaBr$_3$ scintillator on non-proportionality was also performed with the SLYNCI apparatus. Table 1 lists the characteristics of the crystals measured on SLYNCI. Except for the 0.5% cerium dopant level crystal (grown at RMD), all the crystals were manufactured by Saint Gobain and are marketed under the name "BrilLanCe 380". The 5% cerium dopant level represents the standard crystal produced by Saint Gobain. Note that the light yield shows a decreasing trend with increasing cerium concentration. The light yield measurement for the 5% dopant level crystal is based on what was seen in SLYNCI as opposed to a separate measurement. The relative light yield curves for all these crystals are shown in figure 3. The curves are all very similar with the curves being the same at the few percent level, but note there is no significant rise in relative yield as the electron energy decrease from 446 keV. There are also some significant systematic differences, again at higher energy. These differences trend with increasing dopant level with the 0.5% Ce and 5% dopant level showing no difference.

Ce Dopant Level	Manufacturer	Crystal Size	Light Yield (photons/MeV)
0.5%	RMD	~10x10x10mm	77,000
5.0%	St. Gobain	10mmØx10mm	~70,000
10.0%	St. Gobain	10mmØx10mm	71,000
20.0%	St. Gobain	10mmØx10mm	61,000
30.0%	St. Gobain	20mmØx10mm	27,000

Table 1. A list of the LaBr$_3$ crystals used in the non-proportionality study.

DISCUSSION

To further interpret the LaBr$_3$ data, we have fit the data the formulation described in reference [10]:

$$\eta_{CAP} = \{1 - \eta_{ONS} \exp[-(dE/dx) / (dE/dx)_{ONS}]\} / [1 + (dE/dx) / (dE/dx)_{BIRKS}] \qquad (1)$$

which includes the formation of excitons from holes and electrons (maximum efficiency of η_{ONS}) based on the so-called Onsager mechanism of Coulombic attraction, and incorporates the two stopping power parameters that describe the attraction between electrons and holes, $(dE/dx)_{ONS}$, and the annihilation between nearby excitons, $(dE/dx)_{BIRKS}$. The parameters used in Eq. (1) to fit the data are compiled in Table 2, based on integrating Eq. (1) from the initial to final energies. We notice that η_{ONS} value drops from 18.5% to 12% in passing from 0.5% Ce-doping to 30% doping. This means that less excitons are created (after the cascade) as the doping level is increased, probably because the carriers are more likely to be trapped at the Ce ions so that they are unable to recombine on a relevant timescale. We also notice that there is a slight increase in the Birks parameter $(dE/dx)_{BIRKS}$ indicating that exciton-exciton annihilation becomes less probable at high Ce doping. Lastly, the $(dE/dx)_{ONS}$ term is approximately constant as it is predicted to only be dependent on the host's dielectric constant.

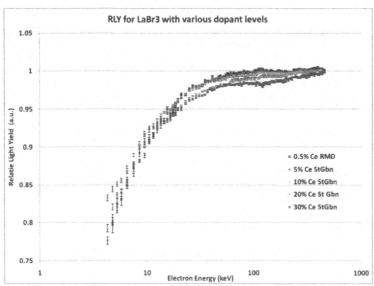

Figure 3. Relative light yield curves for LaBr₃ with various Ce dopant levels.

LaBr₃ Ce doping level	$(dE/dx)_{ONS}$ (MeV/cm)	h_{ONS} (%)	$(dE/dx)_{BIRKS}$ (MeV/cm)
0.5%	36.4	18.5	370
5%	36.4	18	392
10%	36.4	15.5	444
20%	36.4	13	455
30%	36.4	12	455

Table 2. Results of fits of the LaBr₃(Ce) non-proportionality curves to equation 1.

CONCLUSIONS

Non-proportionality studies of various scintillators have been studied on SLYNCI. The non-proportionality of various alkali halides have been measured, showing similar behavior for all the crystals tested, with NaI showing a slightly better non-proportionality. A SrI₂(Eu) crystal has also been tested which a similar shape but less non-proportionality. A series of LaBr₃ crystals have also been tested, looking at the impact of Ce dopant level on non-proportionality. All the LaBr₃ crystals showed very similar behavior with only small systematic difference seen with increasing dopant level. The studies of alkali halide and strontium halide are still a work in progress and more samples will be run in the near future.

ACKNOWLEDGMENTS

This work is supported by the National Nuclear Security Administration, Office of Defense Nuclear Nonproliferation, Office of Nonproliferation Research and Development (NA-22) of the U.S. Department of Energy. Thanks to Eric Mattmann and Bruno Aleonard from St. Gobain for the loan of the LaBr$_3$(Ce), "BrilLanCe 380" crystals. This work performed under the auspices of the U.S. Department of Energy by Lawrence Livermore National Laboratory under Contract DE-AC52-07NA27344 and by Lawrence Berkeley National Laboratory under Contract No. DE-AC02-05CH11231.

REFERENCES

1. W.-S. Choong, K.M. Vetter, W.W. Moses, S.A. Payne, N.J. Cherepy, J.D. Valentine and G. Hull, "Design of a facility for measuring scintillator non-proportionality," *IEEE Trans. Nucl. Sci.*, vol. 55, pp. 1753-1758, 2008.
2. W.-S. Choong, G. Hull, W.W. Moses, K.M. Vetter S.A. Payne, N.J. Cherepy, J.D. Valentine, "Performance of a Facility for Measuring Scintillator Non-Proportionality," *IEEE Trans. Nucl. Sci.*, vol. 55, pp. 1073-1078, 2008.
3. G. Hull, S. Du, T. Niedermayr, S. Payne, N. Cherepy, A. Drobshoff, and L. Fabris "Light collection optimization in scintillator based gamma-ray spectrometers" *Nucl. Instr. Meth. A*, vol. 588, 384-388, 2008.
4. G. G. Kelley, P. R. Bell, R. C. Davis and N. H. Lazar, "Intrinsic scintillator resolution," *IRE Transactions in Nuclear Science*, vol. NS3, pp. 57-58, 1956.
5. C. L. Melcher, "Perspectives on the future of development of new scintillators," *Nuclear Instruments and Methods in Physics Research A*, vol. 537, pp. 6-14, 2005.
6. J. E. Jaffe, "Energy and length scales in scintillator nonproportionality," *Nucl. Instr. & Meth. A, vol.* 580, 1378-1382 (2007); J. E. Jaffe, D. V. Jordan and A. J. Peurrung, "Energy nonlinearity in radiation detection materials: Causes and consequences," *Nuclear Instruments and Methods in Physics Research A*, vol. 570, pp. 72-83, 2007.
7. B.D. Rooney and J.D. Valentine, "Benchmarking the Compton Coincidence Technique for measuring electron response nonproportionality in inorganic scintillators ", IEEE Trans. Nucl. Sci., vol.43, no. 3, pp. 1271-1276, Jun. 1996.
8. T.D. Taulbee, B.D. Rooney, W. Mengesha, J.D. Valentine, "The measured electron response nonproportionalities of CaF2, BGO, and LSO", IEEE Trans. Nucl. Sci., vol. 44, no. 3, pp. 489-493, Jun. 1997
9. W. Mengesha, T.D. Taulbee, B.D. Rooney, J.D. Valentine, "Light yield nonproportionality of CsI(Tl), CsI(Na), and YAP", IEEE Trans. Nucl. Sci., vol. 45, no. 3, pp. 456-461, Jun. 1998.
10. S. A. Payne, et al., *IEEE Trans. Nucl.* accepted for publication.

Materials II

Mater. Res. Soc. Symp. Proc. Vol. 1164 © 2009 Materials Research Society   1164-L08-02

First-principles study of defects and carrier compensation in semiconductor radiation detector materials

Mao-Hua Du, Hiroyuki Takenaka, and David J. Singh
Materials Science & Technology Division and Center for Radiation Detection Materials and Systems, Oak Ridge National Laboratory, Oak Ridge, TN 37831, USA

ABSTRACT

We discuss defect engineering strategies in radiation detector materials. The goal is to increase resistivity by defect-induced Fermi level pinning without causing defect-induced reductions in the carrier drifting length. We show calculated properties of various intrinsic defects and impurities in CdTe. We suggest that the defect complex of a hydrogen atom and an isovalent impurity on an anion site may be an excellent candidate in many semiconductors for Fermi level pinning without carrier trapping.

INTRODUCTION

Semiconductor radiation detectors need to have high resistivity in order to suppress free carrier noise.[1] The compound semiconductors that give large enough band gaps for room temperature operation cannot be sufficiently purified to obtain intrinsic semiconductor behavior in contrast to small-gap elemental semiconductors like germanium. High resistivity in large-gap compound semiconductors approaching the theoretical intrinsic semiconductor limit can be made possible by the presence of midgap impurity or defect levels that pin the Fermi level, as schematically shown in Fig. 1(a). A typical example is semi-insulating GaAs, whose high resistivity is made possible by As antisite (As_{Ga}) defects that induce a midgap energy level.[2] However, the deep levels are also efficient nonradiative recombination centers that greatly reduce the carrier drifting length, the figure of merit for semiconductor radiation detectors. This problem severely limits the applicability of GaAs as a room-temperature radiation detector. The same carrier compensation mechanism, i.e., Fermi level pinning by anion antisites, has been suggested for CdZnTe (CZT),[3, 4] the best room temperature radiation detector available now. What is puzzling is that if the anion antisite is the common Fermi level pinning mechanism for both GaAs and CZT, why does the same limitation on carrier drifting length not affect CZT as severely as GaAs. We note that, in contrast to GaAs, in which the midgap donor related to As_{Ga} has been confirmed by numerous experiments,[2] the midgap donor in CdTe or CZT, supposedly Te_{Cd} by conventional wisdom, has never been unambiguously observed experimentally. A midgap level at $E_c - 0.75$ eV has been found in n-type CdTe by deep level transient spectroscopy[5], and was tentatively assigned to Td_{Cd}[4]. However, this level was observed only after the CdTe was annealed under saturated Cd vapor. As we will show later, this level is very likely a Cd interstitial level. Moreover, the recent thermoelectric effect spectroscopy measurements did not find any midgap donor levels in CZT.[6]

The carrier compensation mechanism shown in Fig. 1(a) is intrinsically problematic, as a large number of deep donors leads to effective electron trapping. We suggest that the compensation scheme shown in Fig. 1(b) should be a better approach. In Fig. 1(b), there are two different defects whose energy lines cross near midgap while these two defects individually do not induce trap levels or at least do not induce deep trap levels. Therefore, these two defects can pin the Fermi level near the midgap whereas not creating efficient carrier traps or recombination centers.

In this paper, we first show that intrinsic defects (including Te_{Cd}) in CdTe may not be responsible for the high resistivity of CdTe. Instead, we suggest that an extrinsic defect, the O_{Te}-H complex, is an excellent candidate for the carrier-compensating defect. We also discuss bismuth doping in CdTe and its role in carrier compensation and carrier trapping.

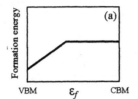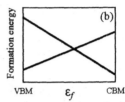

Figure 1. (a) Schematic figure for carrier compensation by a deep donor. The black line is the formation energy of the deep donor defect or impurity as a function of Fermi level. The Fermi level at which the slope of the energy line changes defines the thermal transition level that pins the Fermi level near the midgap. (b) Schematic figure for carrier compensation by two different defects. One is a donor (red line) and the other is an acceptor (blue line). Neither of them inserts a deep trap level. The crossing point of the two energy lines pins the Fermi level near the midgap.

THEORY

We performed calculations based on density functional theory within the local density approximation, as implemented in VASP.[7] The electron-ion interactions are described by projector augmented wave pseudopotentials.[8] The valence wavefunctions are expanded in a plane-wave basis with cutoff energy of 400 eV. All the calculations were preformed using 64-atom cubic cells. A 2×2×2 grid was used for the k-point sampling of Brillouin zone. All the atoms were relaxed to minimize the Feynman-Hellmann forces to below 0.02 eV/Å. The calculated CdTe lattice constant is 6.420 Å, in good agreement with the experimental values of 6.477 Å.[9] Details of the formation energy calculations for defects can be found in Ref. 10.

RESULTS AND DISCUSSION

Intrinsic defects in CdTe

V_{Cd} is usually abundant in CdTe, giving rise to p-type CdTe. It is often assumed that Td_{Cd} compensates holes, leading to high resistivity. Te_{Cd} on a regular Cd lattice site with T_d symmetry is a two-electron deep donor. The $Te_{Cd}(T_d)$-induced gap level is a three-fold degenerate state (t_{2c}) occupied by only two electrons when $Te_{Cd}(T_d)$ is neutral. Consequently, the ground-state structure for the neutral Te_{Cd} is Jahn-Teller distorted with C_{3v} symmetry [see Fig. 2(a)]. The Te_{Cd} is found to be a negative-U center with its (+2/0) transition level only 0.35 eV above the VBM (see Fig. 3). This finding does not support the prevailing assumptions that Te_{Cd}-induced deep donor level pins the Fermi level near the midgap.

The tellurium antisite can also take a stable -2 charge state (Te_{Cd}^{-2}). At this -2 charge state, a large structural relaxation effectively transforms a Te_{Cd}^{-2} to a complex of neutral $(Te\text{-}Te)_{spl}$ and V_{Cd}^{-2} [the $(Te\text{-}Te)_{spl}$ with a missing Cd nearest neighbor, labeled as A in Fig. 2(b)]. Nevertheless, this complex is still labeled as Te_{Cd}^{-2} in this paper. The dissociation of a Te_{Cd}^{-2} to $(Te\text{-}Te)_{spl}$ and V_{Cd}^{-2} costs 0.47 eV according to our calculations. The calculated (0/2-) transition level for Te_{Cd} is close to the midgap as shown in Fig. 3. Although the semi-insulating CdTe samples grown by Bridgman techniques are typically slightly p-type, they could become slightly n-type if the cadmium vacancies are largely suppressed in the high-pressure Bridgman growth or are slightly outnumbered by the intentionally introduced shallow donors. In this case, the Te_{Cd} can act as a deep acceptor to compensate the excess electron carriers.

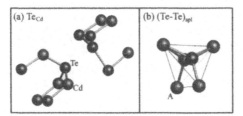

FIGURE 2. Structures of (a) Te_{Cd} at neutral charge states and (b) $(Te\text{-}Te)_{spl}$. The four Cd atoms (vertices of a tetrahedron) around the Te site in (b) are connected by thin grey lines to guide eye.

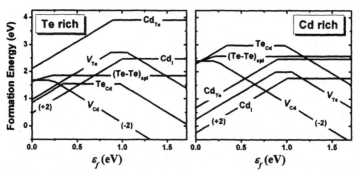

FIGURE 3. Calculated formation energies for various intrinsic defects in CdTe as a function of electron Fermi energy at (a) Te-rich and (b) Cd-rich limits. The slope of an energy line indicates the charge state of the defect, as selectively shown. The transition levels are given by the Fermi energy at which the slope changes.

Cd_i can take two different interstitial sites: one with four Cd neighbors (Cd_i^{Cd}) and the other with four Te neighbors (Cd_i^{Te}). We find that at neutral charge state Cd_i^{Cd} is more stable than Cd_i^{Te}, while at +2 charge state the reverse is true. The +1 charge state is not stable. Thus,

Cd$_i$ is a negative-U center with a deep (+2/0) donor level at E$_c$ – 0.71 eV, near midgap (see Fig. 3). It has been postulated that Cd$_i$ may play a role in the carrier compensation in CdTe radiation detectors.[11] However, the formation energy for Cd$_i$ is relatively high (2.48 eV for neutral Cd$_i$) at the Te-rich limit (typical condition for the growth of single-crystal CdTe radiation detectors).

We have calculated the Cd$_i$ diffusion barriers to be only 0.36 and 0.61 eV at (+2) and neutral charge states, respectively. The semi-insulating CdTe samples used for radiation detectors are typically slightly p-type, favoring the (+2) charge state for Cd$_i$. The small Cd$_i^{2+}$ diffusion barrier of 0.36 eV allows Cd$_i^{2+}$ to diffuse easily at low temperatures, suggesting that Cd$_i$ can be annealed out of the crystal during the slow cooling process. Thus, we expect low Cd$_i$ concentration in the CdTe samples. However, [Cd$_i$] can increase if the CdTe sample undergoes post-growth annealing under saturated Cd pressure. It was reported that such annealing at 800 °C turned the CdTe sample from p-type to n-type and a level at E$_c$ – 0.75 eV was observed,[12] in good agreement with our calculated Cd$_i$-induced level at E$_c$ – 071 eV.

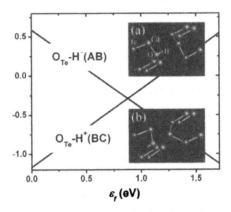

Figure 4. Calculated lowest hydrogenation energies for O$_{Te}$-H complexes at (+), neutral, and (–) charge states. Insets show the structures of (a) O$_{Te}$-H$^+$(BC) and (b) O$_{Te}$-H$^-$(AB) complexes. BC and AB stand for bond-center and anti-bonding positions of hydrogen.

Oxygen-hydrogen complex in CdTe

In intentionally undoped CdTe grown by Brigman techniques, residual oxygen has a concentration typically orders of magnitudes higher than other residual impurities and comparable to intrinsic defects.[13, 14] When H binds O$_{Te}$, the O$_{Te}$-H complex is stable as either a donor or an acceptor depending on the Fermi level, as shown in Fig. 4. The neutral O$_{Te}$-H is unstable. The O$_{Te}$-H (+/–) transition level is calculated to be E$_c$ – 0.73 eV, near the midgap (see Fig. 4). Since the O$_{Te}$-H complex has low energy, it is expected to have appreciable concentration. As an amphoteric defect, the O$_{Te}$-H complex compensates either p-type or n-type carriers, whichever is in excess. Furthermore, both (+/0) and (0/-) transition levels are outside bandgap. The O$_{Te}$-H$^+$ and O$_{Te}$-H$^-$ complexes have distinctly different structures as shown in the insets of Fig. 4. Their structural conversion to one another is prevented by the large energy

110

barrier and thus the crossing point of energy lines in Fig. 4 is not a trap level. Based on these results, we propose that the O_{Te}-H complex may play an important role in the semi-insulating behavior of CdTe. A similar strategy could be extended to many other semiconductors by introducing defect complexes of isovalent impurity (on anion site) and hydrogen.

Bismuth in CdTe

To satisfy the conditions set in Fig. 2(b), one can also introduce a type of impurity that can be stable either on cation site as a donor or on anion site as an acceptor such that this impurity is amphoteric. Several recent experimental works show that the Bi doping at the level of ~10^{17} cm^{-3} results in semi-insulating CdTe.[15, 16] Calculated formation energies (Fig. 5) show that Bi^+_{Cd} donor is stable when Fermi level is low whereas Bi^-_{Te} acceptor become more stable when Fermi level is high. However, Bi_{Cd} can undergo a Jahn-Teller distortion similar to Te_{Cd} [see Fig. 2(a)]. This stabilizes Bi^-_{Cd} when Fermi level is high (Fig. 5). The trapping of two electrons by Bi^+_{Cd} to form Bi^-_{Cd} is essentially a DX transformation.[17] Bi^-_{Cd} can further bind an O_{Te} with a large binding energy of 1.40 eV. The calculated (0/-) level of Bi^-_{Cd} is E_V + 1.2 eV. The binding of Bi^-_{Cd} and O_{Te} lowers the (0/-) level to E_V + 0.8 eV, close to an experimentally observed Bi-related level at E_V + 0.73 eV.[16] The presence of both Bi^+_{Cd} and Bi^-_{Te} can pin the Fermi level near midgap and thus increase the resistivity. However, Bi^+_{Cd} can trap two electrons to form Bi^-_{Cd} via a DX transformation. This severely limits the electron transport in CdTe. Other group-V impurities in CdTe are likely to suffer the same problem.

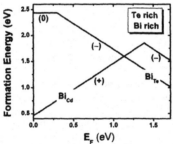

FIGURE 5. Calculated formation energies for Bi_{Cd} and Bi_{Te} in CdTe as a function of Fermi energy at Te-rich limit. The slope of an energy line indicates the charge state of the defect, as shown in the figure. The transition levels are given by the Fermi energy at which the slope changes.

CONCLUSIONS

Using deep donors to increase resistivity in radiation detector materials causes severe problems to electron carrier transport in general. We suggest that a better approach may be to introduce both donor and acceptor defects which have formation energy crossover near the middle of bandgap but individually do not insert trap states in the bandgap. We have investigated the defects and impurities in CdTe. We show that Te_{Cd} donors may not be responsible for the high resistivity of CdTe. Instead, we suggest the the O_{Te}-H complex can pin the Fermi level near the midgap while not introducing trapping levels. A similar strategy can be extended to other semiconductors by introducing defect complexes of isovalent impurity (on anion site) and hydrogen. We further show that Bi doping of CdTe can increase resistivity but limits electron transport due to electron trapping at Bi_{Cd}.

ACKNOWLEDGMENTS

This work was supported by the U.S. DOE Office of Nonproliferation Research and Development NA22.

REFERENCES

[1] T. E. Schlesinger, J. E. Toney, H. Yoon, E. Y. Lee, B. A. Brunett, L. Franks, R. B. James, Mater. Sci. & Eng. **32**, 103 (2001)

[2] G. M. Martin and S. Makram-Ebeid, in *Deep Centers in Semiconductors*, edited by S. T. Pantelides (Gordon and Breach Science Publisher 1992)

[3] M. Fiederle, D. Ebling, C. Eiche, P. Hug, W. Joerger. M. Laasch, R. Schwarz, M. Salk, and K. W. Benz, J. Cryst. Growth **146**, 142 (1995).

[4] M. Fiederle, C. Eiche, M. Salk, R. Schwarz, K. W. Benz, W. Stadler, D. M. Hofmann, and B. K. Meyer, J. Appl. Phys. **84**, 6689 (1998).

[5] G. M. Khattack and C. G. Scott, J. Phys.: Condens. Matter **3**, 8619 (1991).

[6] R. Soundararajan, K. G. Lynn, S. Awadallah, C. Szeles, and S. –H. Wei, J. Electr. Mater. **35**, 1333 (2006).

[7] G. Kresse and J. Furthmüller, Phys. Rev. B **54**, 11169 (1996).

[8] G. Kresse and D. Joubert, Phys. Rev. B **59**, 1758 (1999).

[9] *CRC Handbook of Chemistry and Physics, 88th edition*, D. R. Lide, ed., CRC Press/Taylor and Francis, Boca Raton, FL (2008).

[10] S. B. Zhang, J. Phys.: Condens. Matter **14**, R881 (2002).

[11] R. Grill, P. Hoschl, I. Turkevych, E. Belas, P. Moravec, M. Fiederle, and K. W. Benz, IEEE Trans. Nucl. Sci. **49**, 1270 (2002).

[12] G. M. Khattack and C. G. Scott, J. Phys.: Condens. Matter **3**, 8619 (1991).

[13] N. Krsmanovic, K. G. Lynn, M. H. Weber, R. Tjossem, Th. Gessmann, Cs. Szeles, E. E. Eissler, J. P. Flint, and H. L. Glass, Phys. Rev. B **62**, R16279 (2000)

[14] S. A. Awadalla, A. W. Hunt, K. G. Lynn, H. Glass, C. Szeles, and S. –H Wei, Phys. Rev. B **69**, 075210 (2004).

[15] E. Saucedo, L. Fornaro, N. V. Sochinskii, A. Cuña, V. Corregidor, D. Granados, and E. Diéguez, IEEE Trans. Nucl. Sci. **51**, 3105 (2004).

[16] E. Saucedo, J. Franc, H. Elhadidy, P. Horodysky, C. M. Ruiz, V. Bermúdez, and N. V. Sochinskii, J. Appl. Phys. **103**, 094901 (2008).

[17] P. M. Mooney, J. Appl. Phys. **67**, R1 (1990).

Poster Session

Mater. Res. Soc. Symp. Proc. Vol. 1164 © 2009 Materials Research Society 1164-L09-03

Tb-doped Aluminosilicate Oxyfluoride Scintillating Glass and Glass-ceramic

Z. Pan*[1], K. James[1], Y. Cui[1], A. Burger[1], N. Cherepy[2] and S. A. Payne[2], A. Ueda[1], R. Aga Jr[1], R. Mu[1], and S. H. Morgan[1]

[1]Department of Physics, Fisk University, Nashville, TN 37208, U.S.A.
[2]Lawrence Livermore National Laboratory, Livermore, CA 94550, U.S.A.

ABSTRACT

Two aluminosilicate oxyfluoride glass systems, a lead-cadmium-aluminosilicate oxyfluoride and a lithium-lanthanum-aluminosilicate oxyfluoride, doped with different TbF_3 concentrations, have been fabricated and investigated. By appropriate heat treatment of the as-prepared glasses above, transparent glass-ceramics (TGC) were obtained. The glass-ceramics contain $Tb:Pb(Cd)F_2$ or $Tb:LaF_3$ nano-crystals in the glass-matrix. Differential scanning calorimetry, Raman scattering, and luminescence under both UV and β-particle excitation have been investigated on as-prepared glasses and glass-ceramics. It has been found that the terbium-doped lithium-lanthanum-aluminosilicate oxyfluoride glass exhibits good UV excited luminescence and β-induced luminescence. The luminescence yield increases for glass-ceramic compared to that of the as-prepared glass. The including of LaF_3 in the glass-matrix is beneficial for a higher Tb-doping concentration and a high light yield. The light yield of lithium-lanthanum-aluminosilicate oxyfluoride glass and glass-ceramic is comparable to that of Schott IQI-301 product. However, the terbium-doped lead-cadmium-aluminosilicate oxyfluoride glass and glass-ceramic have a detrimental luminescence performance. The lead cations in the glass-matrix may create non-bridging oxygen defects, which are a strong source of charge traps, and correlated to a strong Raman "Boson" peak.

INTRODUCTION

Glass is attractive scintillating material because of its low-cost, large-volume production possibility, and easy shaping of elements. Glass, however, is a disordered material with significant point defects acting as traps which limit the free carrier transfer from the host matrix to the luminescence centers and provide additional thermal decay channels [1, 2].

Tb-doped glass is slower in response than its Ce-doped counterpart for nuclear radiation detection. But Tb-doped glass has not demonstrated the self-quenching phenomenon which occurs in Ce-doped glass when Ce^{3+} is oxidized to Ce^{4+}. The radiation hardness of Tb-doped scintillating glass is about 2-3 times better than Ce-doped glass [3, 4].

Oxyfluoride glass-ceramics have been developed to combine the particular optical properties of rare-earth ions in a nano-crystalline fluoride host with the chemical stability and mechanical property of oxide glasses. LaF_3 is an excellent host material for rare-earth ions because it has a high solubility for rare-earth ions, low phonon energy (300 - 400 cm^{-1}), and desirable thermal and environmental stability [5, 6].

In this work, two Tb-doped aluminosilicate oxyfluoride glass and glass-ceramic systems were studied. The glass-ceramics were obtained by appropriate heat treatment of the as-prepared glasses. Luminescence under both UV and β-particle excitations have been investigated on as-prepared glasses and glass-ceramics, and compared to that of the Schott IQI-301 scintillating glass product.

EXPERIMENT

The batch compositions of two glasses are $30SiO_2$ -$15(AlO_{1.5})$ -$22PbF_2$ -$29CdF_2$ -$4YF_3$ (LCASOF) and $55SiO_2$ -$6Al_2O_3$ -$28Li_2O$ -$11LaF_3$ (LLASOF). The glasses were fabricated using a conventional melting-quenching technique [7]. The heat-treatment was performed at temperatures between the glass transition temperature T_g and the first crystallization temperature T_c, determined from the results of differential scanning calorimetry (DSC).

The Raman spectra were taken using 514.5 nm laser excitation, on a Spex 1403 double grating spectrometer with a TE cooled PMT and a photon-counting system. PL was measured using 325 nm laser excitation. The incident laser power used is 10 mW. The laser light was focused down to a spot size of about 50 μm in diameter on the sample surface. The luminescence was measured at room temperature.

The β-induced luminescence was acquired using 90Sr/90Y (~1 MeV) source, expected to be equivalent to that produced by gamma excitation. Samples were compared with the Schott IQI-301 glass under the same experimental conditions.

RESULTS AND DISCUSSION

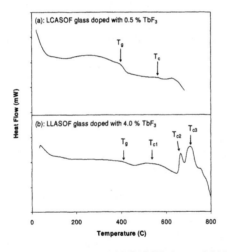

Fig. 1. DSC curves: (a) Tb-doped LCASOF glass and (b) Tb-doped LLASOF glass.

Fig. 1 shows the DSC curves of a Tb-doped LCASOF glass and a Tb-doped LLASOF glass. The LCASOF glass has a T_g at 400 °C and a T_c at 565 °C while the LLASOF has a T_g at 406 °C and multiple T_c's at 540, 667, and 711 °C. The heat-treatment was performed at temperatures

116

between the glass transition temperature and the first crystallization temperature for 6 hours. XRD was performed and reported previously that the glass-ceramics of LCASOF contains Pb(Cd)F$_2$ nano-crystals and LLASOF contains LaF$_3$ nano-crystals in the glass-matrix [7, 8].

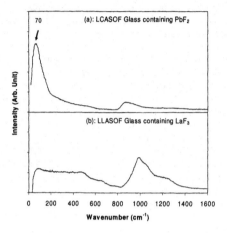

Fig. 2. Raman spectra excited at 514.5 nm: (a) LCASOF glass and (b) LLASOF glass.

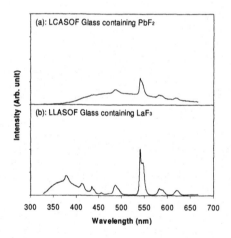

Fig. 3. UV-excited PL of glasses doped with 0.5 at % Tb: (a) LCASOF and (b) LLASOF.

117

Raman spectrum of LCASOF glass in Fig. 2(a) shows a remarkably intense low-frequency Raman peak at 70 cm⁻¹. This so called "Boson" peak has been attributed to the collective modes or cooperative motion of librated elastic assemblies in a weakened glass network [9, 10]. The LCASOF glass contains significant PbF$_2$ component. The lead cations combined with the fluoride anions in glass may cause an increase of the non-bridging oxygen atoms, and these nonbridging oxygen atoms together with loosely bonded Pb^{2+} cations weaken the cohesion of glass network [9]. In Fig. 2(b), the "Boson" peak significantly reduced, implying that LLASOF glass has a more cohesion glass network. LLASOF is lithium-lanthanum-aluminosilicate oxyfluoride, contains no lead cation. It has been reported that constituent aluminum oxide may reduces the number of nonbridging oxygen defects in lithium-aluminosilicate glass matrix [11].

Fig. 3 shows UV-excited PL of (a): a Tb-doped LCASOF glass, and (b): a Tb-doped LLASOF glass, doped with the same Tb concentration of 0.5 at %. Both glasses have significant absorption at the excitation wavelength of 325 nm [8]. The four visible emission bands at 489, 542, 585, and 622 nm are attributed to transitions $^5D_4 \rightarrow {}^7F_i$ (i = 6, 5, 4, and 3) of Tb^{3+} ions. Two significant differences have been observed: (1) the PL intensity of Tb^{3+} in LLASOF is significantly stronger than that of in LCASOF; (2) a broad background emission from 400 to 650 nm is shown in PL of LCASOF, but not in PL of LLASOF. This broad background emission is attributed to defects in glass host and indicates more defects in LCASOF than in LLASOF.

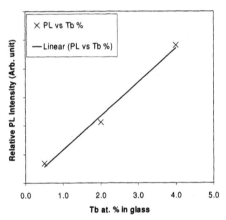

Fig. 4. UV-excited PL intensity versus Tb concentration in LLASOF glass.

The including of LaF$_3$ in the glass-matrix is beneficial for a higher Tb-doping concentration. There is no visual evidence of devitrification for LLASOF doped with 4.0 at %. In contrast, the Tb-doping concentration in LCASOF glass is restricted to be less than 1.0 at % due to spontaneous devitrification. Fig. 4 shows the UV-excited PL intensity versus Tb concentration in LLASOF glass. The PL intensity increases proportionally to the Tb concentration from 0.5 at % to 4.0 at %, indicating no significant self-quenching effect. It has also been observed that the Tb-doped glass-ceramic of LLASOF has a higher luminescence yield compared to that of the as-prepared glass [8].

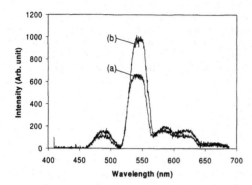

Fig. 5. β-induced luminescence: (a) a LLASOF glass-ceramic doped with 4.0 at % Tb and (b) IQI-301.

Fig. 5 shows β-induced luminescence of a LLASOF glass-ceramic doped with 4.0 at % Tb, compared to that of IQI-301 glass under the same experimental condition. β-induced luminescence shows a similar Tb-concentration effect and heat-treatment effect compared to PL. The β-induced luminescence is strong for Tb-doped LLASOF samples, but extremely weak for Tb-doped LCASOF samples. There are three possible reasons for a significantly higher luminescence yield of Tb-doped LLASOF compared to that of Tb-doped LCASOF: (1) less non-bridging oxygen defects acting as traps to limit the free carrier transfer from the host matrix to the luminescence centers; (2) the constituent ^{6}Li may provide a higher nuclear radiation absorption [11]; (3) a higher doping concentration of Tb^{3+} ions, which is allowed by the high rare earth solubility of the composition LaF_3 in LLASOF host [8].

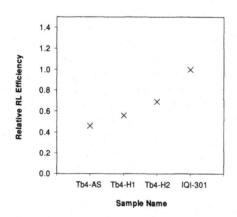

Fig. 6. β-induced luminescence yield of 4 at % Tb doped LLASOF compared with that of IQI-301 glass.

Fig. 6 compares the light yield of four LLASOF samples to that of IQI-301. The four LLASOF samples are all doped with 4.0 at % Tb. Where, Tb4-As: as-prepared, Tb4-H1: heat-treated at 460 °C for 6 h, and Tb4-H2: heat-treated at 480 °C for 6 h. As indicated in Fig. 6, the light yield of Tb-doped LLASOF samples is comparable to that of a scintillating glass IQI-301, which has an even much higher Tb percentage [8]. The light yield of Tb4-H2 is about 71 % of that of IQI-301. This result is very encouraging for further optimization of Tb-doped LLASOF glass and glass-ceramic.

CONCLUSIONS

In summary, we have studied two Tb-doped aluminosilicate oxyfluoride glass and glass-ceramic systems, a lead-cadmium-aluminosilicate oxyfluoride (LCASOF) and a lithium-lanthanum-aluminosilicate oxyfluoride (LLASOF). The including of LaF_3 in the glass-matrix is beneficial for a higher Tb-doping concentration and a higher light yield. The Tb-doped LLASOF glass and glass-ceramic provide a competitive light yield compared to that of IQI-301 scintillating glass product, therefore are promising scintillating material for low-cost, large-volume production. However, the LCASOF glass and glass-ceramic containing PbF_2 showed a detrimental luminescence performance. The detrimental luminescence of LCASOF possibly is related to the increase of non-bridging oxygen because of lead cations, and correlated to a strong Raman "Boson" peak observed.

ACKNOWLEDGMENTS

This research is supported by US National Science Foundation NSF-CREST- CA: HRD-0420516 and NSF-STC CLiPS - grant no. 0423914. The work at LLNL was performed under the auspices of the U.S. DOE by Lawrence Livermore National Laboratory under Contract DE-AC52-07NA27344.

REFERENCES

1. S. Baccaro, A. Cecilia, A. Cemmi, G. Chen, E. Mihokova, and N. Nikl, IEEE Trans. Nucl. Sci. 48, 360 (2001).
2. M. J. Weber, J. Lumin. 100, 35 (2002).
3. G. B. Spector, T. McCollum, and A. R. Spowart, Nucl. Instr. And Meth. A 326, 526 (1993).
4. P. Pavan, G. Zanella, and R. Zannoni, Nucl. Instr. And Meth. B 61, 487 (1991).
5. E. Ma, Z. Hu, Y. Wang, and F. Bao, J. Lumin. 118, 131 (2006)
6. A. C. Yanes, J. Del-Castillo, J. Méndez-Ramos, V. D. Rodríguez, M. E. Torres, and J. Arbiol, Opt. Mater. 29, 999 (2007).
7. Z. Pan, A. Ueda, S. H. Morgan, and R. Mu, J. Rare Earths 24, 699 (2006).
8. Z. Pan, K. James, Y. Cui, A. Burger, N. Cherepy, S. A. Payne, R. Mu, and S. H. Morgan, Nucl. Instr. And Meth. A 594, 215 (2008).
9. Z. Pan, D. O. Henderson, and S. H. Morgan, J. Chem. Phys. 101, 1767 (1994).
10. S. Guha and G. E. Walrafen, J. Chem. Phys. 80, 3807 (1984).
11. M. Bliss, R. A. Craig, and P. L. Reeder, Nucl. Instr. And Meth. A 342, 357 (1994).

Mater. Res. Soc. Symp. Proc. Vol. 1164 © 2009 Materials Research Society   1164-L09-10

Origins of Twinned Microstructures in $B_{12}As_2$ Epilayers Grown on (0001) 6H-SiC and Their Influence on Physical Properties

Yu Zhang[1], Hui Chen[1], Ning Zhang[1], Michael Dudley[1], Yinyan Gong[2], Martin Kuball[2], Zhou Xu[3], Yi Zhang[3], James H. Edgar[3], Lihua Zhang[4], Yimei Zhu[4]

[1]Department of Materials Science and Engineering, Stony Brook University, Stony Brook, NY 11794-2275

[2]H.H. Wills Physics Laboratory, University of Bristol, Bristol, United Kingdom

[3]Department of Chemical Engineering, Kansas State University, Manhattan, KS

[4]Center for Functional Materials, Brookhaven National Laboratory, Upton, NY

ABSTRACT

The defect structure in $B_{12}As_2$ epitaxial layers grown at two different temperatures on (0001) 6H-SiC by chemical vapor deposition (CVD) was studied using synchrotron white beam x-ray topography (SWBXT) and high resolution transmission electron microscopy (HRTEM). The observed differences in microstructures were correlated with the differences in nucleation at the two growth temperatures. The effect of the difference in microstructure on macroscopic properties of the $B_{12}As_2$ was illustrated using the example of thermal conductivity which was measured using the 3-ω technique. The relationship between the measured thermal conductivity and observed microstructures is discussed.

INTRODUCTION

$B_{12}As_2$ is a member of the icosahedral boride family with a structure consisting of twelve boron atom icosahedra arranged at the corners of a rhombohedral unit cell with two-atom As-As chains along the body diagonal. It has a band gap of 3.47eV, and the material has the extraordinary ability to "self-heal" after radiation damage, making it potentially useful for devices operating in high electron radiation environments [1-6]. The absence of native substrates necessitates the heteroepitaxial growth of $B_{12}As_2$, typically on 6H-SiC substrates, often achieved using chemical vapor deposition [7]. Epitaxial growth on (0001) 6H-SiC is facilitated by the fact that the in-plane lattice constants of 6H-SiC are close to one half of those of $B_{12}As_2$. Gaining a detailed understanding of its microstructure is essential on the pathway to device demonstration. In this paper, we present studies of the influence of growth temperature on the microstructure of $B_{12}As_2$ thin films grown on (0001) 6H-SiC substrates. Implications of the microstructure on selected macroscopic physical properties are discussed.

EXPERIMENT

On axis, c-plane 6H-SiC wafers were used as substrates for the CVD growth of $B_{12}As_2$. The $B_{12}As_2$ films were synthesized by employing gaseous precursors of 1% B_2H_6 in H_2 and 2% AsH_3 in H_2. The c-plane $B_{12}As_2$ was deposited at 1275°C and 1450°C, for samples, denoted here S1 and S2, respectively, with 500 Torr of reactor pressure. Non-destructive SWBXT was carried out at the Stony Brook Topography station at the National Synchrotron Light Source, Brookhaven National Laboratory. Following this, cross-sectional TEM samples were made parallel to ($1\bar{1}20$) 6H-SiC. Conventional and high resolution TEM observation was performed using a JEOL 2100 transmission electron microscope with an electron accelerating voltage of 200KeV. To illustrate the implication of the microstructure on macroscopic physical properties of the film, the example of thermal conductivity was considered, measured using the 3-ω technique, which is extensively used to measure thermal conductivities of bulk and thin film dielectric materials [8-9]. An ~100nm thick silicon oxide film was deposited onto the $B_{12}As_2$ film by plasma enhanced CVD to electrically isolate the sample. After this, photolithography was used to define a pattern of metal line heaters followed by the sputter deposition of the metal. An AC voltage of frequency ω was applied to the line heater, resulting in an increase in temperature. Because the resistivity of the metal depends on temperature, the voltage along the line heater also changes. The 3ω component in the resulting AC voltage was measured, which can be used to calculate the thermal conductivity.

RESULTS AND DISCUSSION

Figure 1 shows the indexed diffraction patterns of sample S1 and S2 obtained via SWBXT in transmission using a 1mm^2 area incident white beam. The diffraction spots labeled using the four index system correspond to the 6H-SiC substrate and exhibit six-fold symmetry as expected. For both samples, exposure times were increased in order to allow the weaker $B_{12}As_2$ diffraction spots to accumulate sufficient exposure to enable indexing. The diffraction spots from the $B_{12}As_2$ were generally more diffuse and appeared to form a 6-fold symmetric pattern rather than the 3-fold symmetric pattern expected for the rhombohedral crystal structure of $B_{12}As_2$. Detailed analysis confirms that this is because both films were twinned and, using the three index system, subscripts I and II are used to indicate matrix and twin diffraction spots. The size and shape of the matrix and twin diffraction spots indicate that the films appear to be fairly homogeneous solid solutions of matrix and twin.

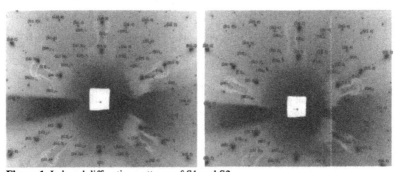

Figure 1. Indexed diffraction patterns of S1 and S2.

The presence of rotational variants in the form of twins is expected in such degenerate epitaxy, as a lower symmetry film is epitaxially grown on a higher symmetry substrate, and the rotational variants are related by the symmetry element present in the substrate which is absent in the film. [10-13]. The details of the microstructure of the twin domains are, however, expected to vary with growth temperature. In order to explore this variation in more detail, both samples were studied using TEM and conventional TEM images recorded from the samples are shown in Figures 2-4. In sample S1, sets of long, parallel (111) twin boundaries are confined to a region ~200nm thick comprising the volume from the substrate/epilayer interface to about one half of the thickness of the epilayer. The average thickness of the twin domains separated by the neighboring twin boundaries is around 60nm (see schematic diagram of the microstructure in Figure 2(c)). In sample S2, twin boundaries (lateral in this case) are mostly perpendicular to the epilayer interface, and the average distance between neighboring lateral twin boundaries (width of one domain) is around 140nm. (See schematic diagram of the microstructure in Figure 2(d))

Enlarged HRTEM images recorded from samples S1 and S2 are shown in Figures 3 and 4, respectively. Figure 3 shows a (111) twin boundary separating two twinned domains. This and other such boundaries have slightly increased complexity in their images since not only are they stepped parallel to the interface on these images but also into the plane of the page. Figure 4 shows a fairly complex microstructure comprising an almost columnar domain configuration suggesting that it may originate due to multiple nucleation. Note that the interface in both samples exhibits steps of opposite sign as expected for these "on-axis" substrates.

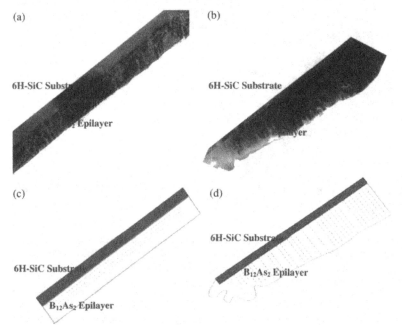

Figure 2. Low magnification TEM images of epilayers in S1 ((a)) and S2 ((b)). Note the boundaries between twinned domains approximately parallel to the interface in the interface region in (a) and penetrating the epilayer perpendicular to the interface in (b). Schematic diagram of epilyaers in S1((c)) and S2 ((d)). Dotted lines indicate twin boundaries.

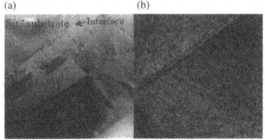

Figure 3. Enlarged TEM image of the twin boundaries parallel to the interface in sample S1 ((a) and HRTEM image of one of the (111) boundaries ((b)).

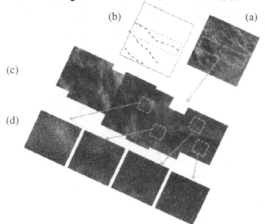

Figure 4. Enlarged TEM image of the lateral twin boundaries in sample S2 ((a) and (c)) and schematic diagram ((b)). HRTEM images of the indicated regions both matrix and twin domains ((d)).

Some understanding of the origin of these quite different microstructures can be obtained by consideration of the different nucleation under the two sets of growth conditions. Most growth parameters are similar except for the growth temperature which is 1275 °C for S1 and 1450 °C for S2. One might intuitively expect that the lower growth temperature would create a higher super saturation and thus a higher nucleation rate. However, contrary to expectation, the sample which appears to exhibit a microstructure resulting from multiple nucleation effects is sample S2 which was grown at the higher temperature. Insight into this can be provided by consideration of the species likely involved in the CVD growth. For the 1% B_2H_6 and 2% AsH_3 in H_2 precursors utilized one might expect gas phase reaction at the growth temperature to involve species such as icosahedral $B_{12}H_{12}$ reacting with arsine to form small subsets of the $B_{12}As_2$ unit cell [13]. Given the relatively large size of these resultant species, it seems reasonable to assume that their surface diffusion will be complex and that they might be subject to the "lock-and-key" effect valid for the surface diffusion of large organic molecules on metal substrates [14]. For such systems only those molecules adsorbed onto the surface that are not oriented correctly to form bonds to the surface have the opportunity to become mobile and diffuse. If they are oriented and positioned to form bonds (multiple bonds) they are locked onto the surface and become immobile. Thus, in such systems, surface diffusion is a greater barrier to the nucleation process than thermodynamics. For the growth of $B_{12}As_2$ epilayers grown on on-axis (0001) 6H-SiC substrates, species can bond to the substrate in either "matrix" or "twin" orientation [12] and species are most effectively "locked-in" if they are able to bond simultaneously to a terrace and a step riser [12-13].

For sample #S1, the lower growth temperature of 1275°C leads to higher super-saturation but also more difficult surface diffusion compounded by the large size of the species. Thus, it is expected that more species will drop onto the surface but once there they will not undergo much surface diffusion. Only those that happen to drop in the right orientation close to a step riser will be locked in by bonding to both terrace and riser with all other being expected to desorb. Those that lock in will develop into nuclei and expand sideways until they impinge upon adjacent nuclei which may be in matrix or twin orientation. Since, the probability of such fortuitous adsorption is relatively low, this will lead to just a few nuclei being formed which will then coalesce and over grow each other to form a relatively small number of domains separated by twin boundaries roughly parallel to the interface consistent with the microstructure observed in sample S1.

For sample #S2, the higher growth temperature of 1450°C leads to lower super-saturation but easier surface diffusion which means that while fewer species will drop onto the surface but those that do are able to experience more surface diffusion. Some species, as at the lower temperature, may by chance be correctly positioned and oriented to immediately bond to terrace and riser while the majority will have the chance to diffuse around until they find a step, get locked in and form a nucleus (still others who don't find a step will desorb). Therefore, in this case, in contrast to the lower temperature case, it is more likely that we will have multiple nucleation and smaller domains leading to a larger number of nuclei which impinge upon each other at an earlier stage of growth which will be separated by many complex twin domain walls roughly perpendicular to the interface. This is consistent with the microstructure observed in sample S2.

To illustrate the implication microstructures have on macroscopic physical properties of the $B_{12}As_2$, preliminary measurements of the thermal conductivity of the $B_{12}As_2$ along the $[111]_{B_{12}As_2}$ direction, i.e., perpendicular to the film surface, were performed using the 3-ω technique. To measure thermal conductivity of thin film, one-dimension heat conduction model and semi-infinite substrate are assumed. Qualitatively, it has been estimated that the errors related to these two factors are within 1% when the thermal penetration depth is at least 5 times greater than the width of the line heater and smaller than one-fifth of the sample thickness. Both conditions are fully satisfied in the frequency range used in our experiments [15]. We determined at room temperature a thermal conductivity of 0.148 $Wcm^{-1}K^{-1}$ for sample S1 and 0.256 $Wcm^{-1}K^{-1}$ for sample S2. This difference is in part due to the difference in microstructure for both samples, with S1 having a larger density of grain boundaries/domain walls than S2 parallel to the sample surface. Thermal conductivity is affected by phonon scattering occurring at these twin or grain boundaries and will diminish thermal conductivity more for sample S1 than S2. One of the potential application of IBA is for thermoelectrics. The conversion efficiency from heat to electricity depends on the figure-of-merit, which is inverse proportional to thermal

125

conductivity, and proportional to product of the square of thermal power and electrical conductivity. For this purpose, we want material with low thermal conductivity. The HRTEM results help to understand how to manipulate the thermal conductivity by varying growth condition. People have been working on develop complex material with low thermal conductivity and high thermal power. Further contributions to difference in thermal conductivity may arise from differences in the impurity concentration. A detailed analysis will be reported elsewhere [16].

CONCLUSIONS

The defect structures in $B_{12}As_2$ epitaxial layers grown at two different temperatures on (0001) 6H-SiC by chemical vapor deposition (CVD) was studied using synchrotron white beam x-ray topography (SWBXT) and high resolution transmission electron microscopy (HRTEM). Differences in the observed microstructures were correlated with difference in nucleation at different growth temperature arising from the "lock-and-key" effect. We show that macroscopic physical properties of the $B_{12}As_2$ are affected by the microstructure, as illustrated by the example of thermal conductivity.

ACKNOWLEDGEMENT

The authors gratefully acknowledge support by the National Science Foundation (Materials World Network) under Grant No.0602875 and by the Engineering and Physical Science Research Council (EPSRC) under Grant No. EP/D075033/1 under the NSF-EPSRC Joint Materials Program. The X-ray topography experiments have been carried out at Stony Brook Topography Facility (Beamline X19C) at the National Synchrotron Light Source (NSLS), Brookhaven National Laboratory (BNL), which is supported by the U.S. Department of Energy (D.O.E.) under Grant No. DE-AC02-76CH00016. Research carried out in part at the Center for Functional Nanomaterials, BNL, which is supported by the U.S. D.O.E., Division of Materials Sciences and Division of Chemical Sciences, under contract No. DE-AC02-98CH10886. We thank J. Gray (University of Bristol) for contributions to this work.

REFERENCES

1. D. Emin, *Physics Today*, **55**, January (1987).
2. D. Emin, *J. Sol. Sta. Chem.*, **179**, 2791 (2006).
3. D. Emin, T. L. Aselage, *J. App. Phys.*, **97**, 013529 (2005).
4. D. Emin, *J. Sol. Sta. Chem.*, **177**, 1619 (2004).
5. M. Carrard, D. Emin, L. Zuppiroli, *Phys. Rev. B*, **51**(17), 11270 (1995).
6. T.L. Aselage and D. Emin, Boron Carbides, in CRC Handbook of Thermoelectrics, D.M. Rowe (Ed.), CRC Press, Boca Raton, (1995).
7. R. Nagarajan, Z. Xu, J. H. Edgar, F. Baig, J. Chaudhuri, Z. Rek, E. A. Payzant, H. M. Meyer, J. Pomeroy and M. Kuball, *J. Crystal Growth.*, **273**, 431 (2005).
8. Cahill and Pohl, *Phys. Rev.*, B **35**, 4067 (1987)
9. Cahill et al., *Rhys. Rev.*, B **50**, 6077 (1994)
10. S. W. Chan, *J. Phys. Chem. Solids*, **55**, 1137 (1994).
11. C. P. Flynn and J. A. Eades, *Thin Solid Films*, **389**, 116 (2001).
12. H. Chen, G. Wang, M. Dudley, L. Zhang, L. Wu, Y. Zhu, Z. Xu, J. H. Edgar, and M. Kuball, *J. Appl. Phys.*, 103 (12), 123508 (2008)
13. Hui Chen, Guan Wang, Michael Dudley, Zhou Xu, J. H. Edgar, Tim Batten, Martin Kuball,

Lihua Zhang, and Yimei Zhu, *Appl. Phys. Lett.*, 92 (23), 231917 (2008)
14. Roberto Otero, Frauke Hümmelink, Fernando Sato, Sergio B. Legoas, Peter Thostrup, Erik Lægsgaard, Ivan Stensgaard, Douglas S. Galvão and Flemming Besenbacher, *Nature Materials*, **3**, 779 - 782 (2004)
15. T. Borca-Tasciuc et al., *Review of Scientific Instruments*, 72, 2139 (2001)
16. Y. Gong, J. Gary, H. Chen, Yu Zhang, M. Dudley, Yi Zhang, J. Edgar, and M. Kuball, (unpublished).

Mater. Res. Soc. Symp. Proc. Vol. 1164 © 2009 Materials Research Society   1164-L09-11

Structural, optical and electrical characterization of thermally evaporated TlBr thin films

Natália Destefano and Marcelo Mulato
Department of Physics and Mathematics, School of Science, Philosophy and Letters of Ribeirão Preto, University of São Paulo, Av. Bandeirantes 3900, 14040-901, Ribeirão Preto-SP, Brazil

ABSTRACT

This paper presents the study related to the production of TlBr thin films. Films produced by thermal evaporation present better structural properties than those produced by spray pyrolysis. The main XRD peak of the evaporated films correspond to the (100) crystalline plane. The structure is columnar as revealed by cross section SEM. The thickness decreases with increasing deposition height. Optical band gap of 3.0 eV and electrical resistivity of about 10^9 Ωcm were obtained. EDS reveals a reduction in the amount of Br in the final films. One order of magnitude was obtained for the photo-to-dark current ratio when irradiation in the medical diagnosis X-ray mammography energy range was used.

INTRODUCTION

The properties of high atomic number and high band gap compound semiconductors for X-ray and γ-ray detection have been investigated over many years. Crystalline material obtained as bulk crystals and, more recently, polycrystalline thick semiconductor films have been evaluated for this intent [1-3].

Due to its high atomic number (Tl = 81 and Br = 35), high mass density (7,56 g/cm^3) and intrinsic band gap (2.68 eV), thallium bromide (TlBr) is a very promising semiconductor candidate for room temperature radiation detectors [4-6]. In previous papers extensive studies on the purity of the TlBr starting powder and single crystal growth (bulk) have been reported by several authors [8-14]. However, there are few works related to the study of this material in the thin film polycrystalline form for applications in large area medical diagnosis equipments [15-16]. In this work, spray pyrolysis and thermal evaporation were used as alternative methods for the deposition of TlBr polycrystalline films. Both methods present relative low cost and can be easily expanded for large areas. The aim of this work is to compare the different techniques and to study the influence of the main growth conditions on the final properties of TlBr films (structural, optical and electrical).

EXPERIMENT

Mili-Q water was used as solvent for the production of films by spray pyrolysis. The solution (0,10g of TlBr dissolved in 100g of water) was stirred at 70°C. Each deposition run used 1 cm x 1 cm substrates at 100°C, on top of which the solution was sprayed inside a positive pressure chamber. Nitrogen was the carrying gas at a rate of 8 l/min. The solution flow was 1/90 ml/s, for a nozzle-spray to substrate distance of 19cm. Films produced by thermal evaporation were based on 0.35g of the starting powder that sits on a tungsten crucible. The home-made thermal evaporator is illustrated in Figure 1. The same substrate sizes were used, and they were held at room temperature in this case. The separation between evaporation boat and substrates, from here on called deposition height, was varied from 3 up to 9 cm. Films with thickness varying from 2 to 30 μm were obtained for growth rates (which were not tightly controlled but estimated after each deposition) of 0.4 to 6.0 μm/min. The boat current was about 58A, what lead to boat temperatures below 400°C, according to the previous calibration of the system.

Figure 1: Photography of home-made thermal evaporator.

Corning glass 7059 and quartz were used as substrates. For the electric transport measurements, carbon contacts were applied onto the samples surface. The structure of the films was investigated by X-ray Diffraction (XRD), the morphology by Scanning Electron Microscopy (SEM) and the composition by Energy Dispersive X-Ray Spectroscopy (EDS). Optical transmittance as a function of wavelength were performed to estimate the optical gap of the TlBr films. Resistivities were also measured using current versus voltage experiments. The dark current was compared to the current of the films under irradiation of a medical diagnosis X-ray system, in the mammography energy range.

RESULTS AND DISCUSSION

XRD data (Figure 2) reveals that the evaporated films present better structural ordering when compared to the sprayed ones. For all evaporated TlBr films we observed the presence of two main peaks corresponding to the Miller indices of reflections 100 and 200. Nevertheless the result varies as a function of deposition height: the ratio of the integrated areas of the two peaks increases exponentially, as presented in Figure 3. This variation might possibly be related to the presence of more than a single phase inside the material. The average grain size of the TlBr films was found to be 80 nm.

SEM pictures (see figure 4(a)) reveal that the sprayed films present a large number of holes with varying sizes, leading to a poor coverage of the substrate. These holes might be due to the regions where the excess of solvent accumulates prior to evaporation from the surface of the substrate, thus preventing the deposition of the material in these areas. In order to avoid this problem a slower deposition rate should be used but this would represent too much longer times for the specific industrial applications. A tilted image presented in figure 4(b) shows that it is practically impossible to determine the thickness of the film, due to its large non-homogeneity.

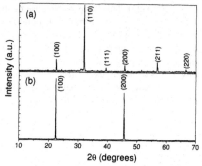

Figure 2: XRD data for films produced by: a) spray pyrolysis and b) thermal evaporation.

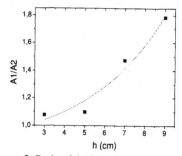

Figure 3: Ratio of the integrated areas of peaks (100) – A1 and (200) – A2 as a function of deposition height.

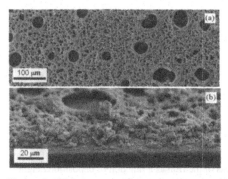

Figure 4: SEM pictures of films produced by spray pyrolysis: (a) surface and (b) tilted.

For the thermally evaporated films, cross-section SEM photographs (Figure 5(a)) reveal a columnar structure. Furthermore, the results show that with increasing deposition height a better coverage is obtained (see figure 5(b) and 5(c), respectively). This happens because the higher deposition height leads to a smaller deposition rate, with larger time for redistribution of the ad-atoms during deposition. We verified an exponential decrease for the thickness (obtained from cross section SEM) as a function of the deposition height (Figure 6(a)). The evaporation of the material occurs in a spherical form and the amount that arrives at the substrate decays exponentially with increasing distance. Moreover, this behavior also reflects a possible change in the deposition rate for different heights. EDS results show that the thermal evaporation technique influence the final composition of the films, with a Br loss of about 10% as compared to the original powder.

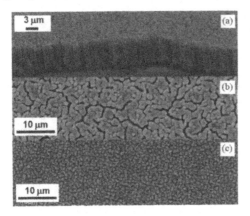

Figure 5: SEM pictures of films produced by thermal evaporation: (a) cross section; (b) and (c) surface. (a) and (b) correspond to a deposition height of 5 cm, while (c) corresponds to 9 cm.

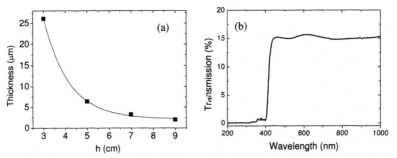

Figure 6: (a) Exponential decay of the thickness of the film as a function of deposition height (h). (b) Optical transmission as a function of wavelength for thermally evaporated films.

Optical transmission as a function of wavelength (Figure 6(b)) shows a pronounced edge at approximately 450 nm, thus leading to an optical gap close to the expected value of 2.68 eV. A small variation with increasing deposition height is observed, possibly indicating the existence of a material with a smaller amount of defects inside the band gap. Note that in our case the larger optical gap was about 3.0 eV, for the larger deposition height of 9 cm. The larger value of band gap when compared to the literature might also be due to the variation of the amount of Br in the alloy as already discussed.

Electrical resistivities in the dark lies close to 1.12×10^9 Ωcm. This value is similar to others reported in the literature [15, 17]. Nevertheless current instabilities were observed as in reference 15. In our case, for an electric field of 1.3×10^{-3} V/μm a standard deviation in current of about 6% was observed.

Finally, the dark current was compared to the current of the films under irradiation using medical diagnosis X-ray, in the mammography energy range. The X-ray generator was operated at 30kVp and 50mAs. The results in figure 7(a) correspond to a linear behavior of the current even for voltages up to 50V. More than one order of magnitude is obtained for the corresponding current ratio. The response of the film is also linear as a function of the accelerating voltage of the mammography system as presented in Figure 7(b).

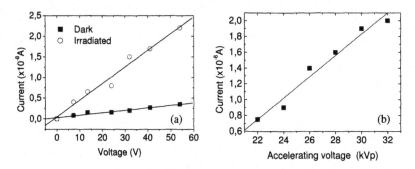

Figure 7: a) Irradiation of the thin film under mammography energy system for varying biases; b) response of the films for varying accelerating voltages of the X-ray tube.

CONCLUSIONS

In summary, two low cost deposition techniques were used to fabricate TlBr thin films aimed for medical applications that require large areas. Thermal evaporation had much better results than spray pyrolysis. The thermally evaporated films presented 10% less Br than the original powder, a lightly larger optical band gap than expected (3.0 eV), and electrical resistivities of about 1×10^9 Ωcm. When illuminated under X-ray mammography energy range, the films presented one order of magnitude photo-to-dark current ratio and a linear response as a function of both, voltage bias and accelerating voltage of the tube, respectively. Thus, the film is a promising candidate for applications in medicine that require large area X-ray photodetectors.

ACKNOWLEDGMENTS

This work was funded by FAPESP (09/00536-2 and 07/015443) and CNPq. The authors thank J. L. Aziani, M. Mano Jr, E. A. Navas, and S.O.B. da Silva for experimental help.

REFERENCES

1. P.J. Sellin, Nuclear Instruments and Methods in Physics Research A 563, 1, 2006.
2. P.J. Sellin, Nuclear Instruments and Methods in Physics Research A 513, 332, 2003.
3. A. Owens and A. Peacock, Nuclear Instruments and Methods in Physics Research A 531, 18. 2004.
4. K. Hitomi, O. Muroi, M. Matsumoto, R. Hirabuki, T. Shoji, T. Suehiro, Y. Hiratate, Nuclear Instruments and Methods in Physics Research A 458, 365-369, 2001.
5. T. Onodera, K. Hitomi, T. Shoji, Nuclear Instruments and Methods in Physics Research A 568, 433–436, 2006.
6. T. Onodera, K. Hitomi, T. Shoji, Y. Hiratate, H. Kitaguchi, IEEE Transactions on Nuclear Science 52, 1999-2002, 2005.
7. K. Hitomi, M. Matsumoto, O. Muroi, T. Shoji, and Y. Hiratate, J. Crystal Growth, 225, 129–133, 2001.
8. K. Hitomi, 0. Muroi, M. Matsumoto, R. Hirabuki, T. Shoji, Y. Hiratate, IEEE Transactions on Nuclear Science 47,777-779, 2000.
9. Alan Owens, M. Bavdaz, G. Brammertz, V. Gostilo, H. Graafsma, A. Kozorezov, M. Krumrey, I. Lisjutin, A. Peacock, A. Puig, H. Sipila, S. Zatoloka, Nuclear Instruments and Methods in Physics Research A 497 , 370–380, 2003.
10. Fábio E. da Costa, Paulo R. Rela, Icimone B. de Oliveira, Maria C. C. Pereira, Margarida M. Hamada, IEEE Transactions on Nuclear Science 53, 1403-1407, 2006.
11. A. Owens, M. Bavdaz, G. Brammertz, V. Gostilo, N. Haack, A. Kozorezov, I. Lisjutin, A. Peacock, S. Zatoloka, Nuclear Instruments and Methods in Physics Research A 497, 359–369, 2003.
12. T. Onodera, K. Hitomi, T. Shoji, Y. Hiratate, Nuclear Instruments and Methods in Physics Research A 525, 199–204, 2004.
13. K. Hitomi, O. Muroi, T. Shoji, Y. Hiratate, H. Ishibashi, M. Ishii, Nuclear Instruments and Methods in Physics Research A 448, 571-575, 2000.
14. K. Hitomi, M. Matsumoto, O. Muroi, T. Shoji, Y. Hiratate, IEEE Transactions on Nuclear Science 49, 2526-2529, 2002.
15. P.R. Bennett, K.S. Shah, L.J. Cirignano, M.B. Klugerman, L.P. Moy, F. Olschner, M.R. Squillante, IEEE Transactions on Nuclear Science, 46, 266-270, 1998.
16. J.P. Ponpon, Nuclear Instruments and Methods in Physics Research A 551 25–26, 2005.

CdTe and CdZnTe Detectors III

Mater. Res. Soc. Symp. Proc. Vol. 1164 © 2009 Materials Research Society   1164-L10-02

A Study of Bismuth Induced Levels in CZT Using TEES/TSC

Raji Soundararajan, Kelly A. Jones, Santosh Swain and Kelvin G. Lynn

Center for Materials Research, Washington State University, Pullman, WA 99164 USA

ABSTRACT

Indium doped and Bismuth co-doped indium doped CdZnTe ingots were grown using a modified vertical Bridgman method. Bismuth co-doping along with Indium doping was aimed at improving the performance of the crystals as room temperature gamma ray detectors. The Indium doped crystals tested were very good radiation detectors with an average bulk resistivity of 2.79×10^{10} Ωcm and $\mu\tau_e$ of 1.52×10^{-3} cm^2/V respectively. While the Bi co-doped Indium doped crystals had a high bulk resistivity ($> 10^{10}$ Ωcm), they were poor radiation detectors. Thermal electric effect spectroscopy (TEES) and thermally stimulated current (TSC) experiments were performed on the Indium doped and the Bi co-doped Indium doped CdZnTe samples to study the Bi related defect levels. The thermal ionization energy and the trap cross-section of the various prominent defect levels were calculated using the variable heating rate TSC experiments. These results have helped determine certain deep defect levels to be uniquely associated to the Bismuth co-doped In doped crystals. At this point, there is not sufficient information to associate the various observed defect levels with specific defects or defect complexes and to identify their nature as donor or acceptor levels. Nevertheless, from TEES data, a p-type conductivity at room temperature has been established in the In doped as well as the Bi co-doped Indium doped samples which indicates holes are the majority charge carriers. Also, it is suggested the deep level B10 with a thermal ionization energy of (0.82 ± 0.02) eV and a large trap cross-section of $(2.59 \pm 5) \times 10^{-12}$ cm^2 served an effective trap center for carrier charges, and in a major way contributed to the deterioration in the performance of the Bi co-doped In doped crystals as effective radiation detectors.

INTRODUCTION

In CdTe and CdZnTe crystals grown via a modified Bridgman method, various intrinsic defects and defect complexes related to V_{Cd} and Te-antisite, acting as trapping and recombination centers are known to be the contributing factors in deteriorating their electrical charge transport properties [1-5]. A successful and preferred method to improving the detector capabilities in these materials has been the implementation of shallow donor group III or VII dopants during growth. The incorporation of shallow dopants facilitates the desired compensation scheme by reducing (though not eliminating) some of these detrimental defect levels and defect complexes in the band gap of the semi-insulating crystals. Hence, selection of an appropriate dopant is of vital importance in producing detector grade crystals. In this regard, industrial practice of growing semi-insulating CdZnTe by doping with chlorine [6] and Indium [7] is a favored industrial practice.

Another technique in which limited work has been done is doping with a heavy element to obtain semi-insulating CdTe and CdZnTe with low carrier trapping. The desirable properties of heavy metals such as (i) low diffusivity of heavy metal atoms in the host CdTe lattice, (ii) a similarity in electronegativity and (iii) ionic radius to Cd atoms facilitates the compensation of native point defects like V_{Cd} with low distortion of the CdTe structure. This implies that doping CdTe with heavy metals in general would not introduce significant amounts of imperfections and would not significantly affect the charge transport properties of CdTe crystals. Studies on CdTe doped with heavy metals like Hg, Ti, Pb, Bi [8-12] have shown increased electrical resistivity compared to pure CdTe crystals. There is a general consensus among the various Bi:CdTe studies regarding the dependence of the conductivity properties of the crystal on the bismuth doping concentration and the amphoteric nature of Bismuth [10-15].

Considering the desirable properties of doping CdTe with heavy metals, this idea was extended to growing Bi co-doped Indium doped CdZnTe with an aim at achieving higher electrical resistivity and better charge transport properties. The Bismuth related defect levels were studied using thermoelectric effect spectroscopy (TEES) and thermally stimulated current (TSC) techniques and the performance of only Indium doped crystals and Bismuth co-doped Indium doped crystals as room temperature radiation detectors were evaluated.

EXPERIMENT

Three different CdZnTe ingots were grown using the vertical Bridgman technique. Ingot I1 was Indium doped $Cd_{0.9}Zn_{0.1}Te$ with no excess Tellurium added during charge preparation. Ingot I2 was Indium doped $Cd_{0.9}Zn_{0.1}Te$ with ~1% excess tellurium added during charge preparation. Ingot B was a bismuth co-doped Indium doped ingot, prepared with 1% excess tellurium. A typical growth was performed under 13 days with an imposed gradient of $60^{0}C$/inch and average rate of 0.8mm/hour.

The as grown Ingots were sand blasted to study the grain structure from the exterior. A few small grains at the tip of the ingot were commonly observed with larger grains growing from the shoulder to the heel in ingots I1 and I2. In ingot B (bismuth co-doped Indium doped ingot) there was heavy nucleation at the tip as shown in Figure 1a, due to constitutional super cooling. Larger grains developed towards the heel during growth as shown in Figure 1b.

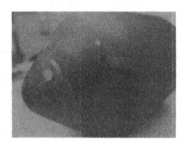

a   b

Figure 1. a. Ingot B after growth a. with large nucleation at the tip; b. larger grain growth beyond the shoulder to the heel

A slice along the growth direction was cut from each ingot. Figure 2a presents a slice cut along the growth direction from Ingot I1 and figure 2b presents a similarly cut slice from ingot I2.

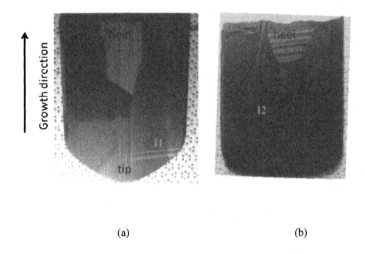

(a)   (b)

Figure 2. a. Vertical slice cut from ingot I1; b. Vertical slice cut from ingot I2.

A conical tip GE224 quartz ampoule was used to load the charge for the growth of ingot I1 and a flat bottomed GE 224 quartz ampoule was used to load the charge for the growth of ingot I2; hence, there was a difference in the shape of the tip of the ingots as can be seen in the figures. Both the slices had large grains, although slice from Ingot I2 showed heavy twinning. Sample I1 with dimensions of ~3 mm x 10 mm x 10 mm was cut from the shoulder of the slice from ingot I1 and sample I2 of similar dimensions was cut close to the mid of the slice from ingot I2 as presented in Figures 2a and 2b.

Figure 3 presents a slice cut along the length of the ingot B. It had large grains and twins. Two single grain and twin free regions were selected from this slice. Samples C (close to the tip of the ingot B) and F (close to the mid of the ingot B) measuring about 3 mm x 10mm x 10mm were cut from this slice.

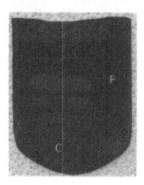

Figure 3. A slice cut along the length of ingot B with regions marking from where sample C and sample F were cut.

All the samples were polished with alumina suspension solutions down to 0.3 micron on all the sides and faces. The final dimensions of the samples were approximately 9mm x 9mm x 2.0 mm. The two (9mm x 9mm) faces were thoroughly cleaned with ethanol and sputtered with gold for electrodes.

A 60 element glow discharge mass spectrometric analysis (GDMS) was performed on bars cut from all the three ingots I1, I2, and B at regions close to where the individual crystals I1, I2, C and F were obtained from. The GDMS results for the samples I1, I2, C and F are presented in Table 1.

Elements (ppb atomic)	Bi	In(dopant)	Na	Al	Fe	Cr	Ni	Cl	Si	Mg	O	Zn %
Sample I1 (Indium doped with no excess Te)	<0.7	5400	180	11	<5	ND	<3	15	6	44	100	4.3%
Sample I2 (Indium doped with 1% excess Te)	<1	2800	79	ND	22	<4	<4	13	220	28	730	3.7%
Sample C (Bi co-doped Indium doped; 1% excess Te) Tip of Ingot B	28	2100	11	150	ND	ND	ND	6	7	17	65	3.3%
Sample F (Bi co-doped Indium doped; 1% excess Te) Mid of ingot B	37	2300	10	160	<5	<5	<6	45	17	25	130	6.0%

Table 1. GDMS data for samples I1 (Indium doped with no excess Tellurium), I2 (Indium doped with 1% excess Tellurium), C and F (Bismuth co-doped Indium doped with 1% excess Tellurium).

The bulk resistivity of the sample was calculated from the current-voltage data, typically from -1V to +1V applied across the sample. A Keithley 2176 with a 6105 resistivity adaptor controlled by a lab-view program was used to record the currents at various bias voltages applied across the sample.

The crystal's response to incident gamma radiation from a ^{57}Co source was studied using a multichannel analyzer. A ^{137}Cs source was used to irradiate the samples to aid in the measurement of the $\mu\tau_e$ product.

Defect levels in the band gap of samples I1, I2, C and F were studied using TEES and TSC measurements. In a typical TEES experiment, the sample was cooled down to about 17K (this was the minimum temperature the system could attain) and illuminated with a suitable IR diode for a 1000 sec. The sample was then heated to about 320K at a constant temperature gradient of 3K maintained across the sample. The applied temperature gradient at the top of the sample was always at 3K higher than the bottom of the sample. Two independent heaters (controlled by a Lakeshore Model 332 temperature controller), one to control the temperature of the top and the other to control the temperature of the bottom of the sample, were used. These experiments were done at low heating rates typically from 0.06K/sec to 0.076K/sec. In a TSC experiment, in addition to a constant 3K gradient, a 10mV to 100mV bias was applied across the sample.

Data from TEES/TSC experiments were plots between the current and average temperature across the sample. Each prominent current peak observed corresponds to a defect level in the band gap of the sample with a definite value for the thermal ionization energy and trapping cross-section.

The thermal ionization energy (E) and capture cross-section (σ) of the observed defect levels were calculated using the variable heating rate method. In this method, the entire Gaussian peak (current peak observed in the TEES or TSC plot) was used. Also, a negligible re-trapping of the liberated charge carriers and a high recombination rate that was a slowly varying function of temperature was assumed [16]. The thermal ionization energy (E_{Th}) of the trap was calculated using Equation 1:

$$E_{th} = k_B T_m \ln\left(\frac{N_c k_B T_m^2 v\sigma}{\beta E_{th}}\right)$$

(1)

which can be rewritten as:

$$\ln\left(\frac{T_m^2}{\beta}\right) = \left(\frac{E_{th}}{k_B T_m}\right) - \ln\left(\frac{k_B N_c v\sigma}{E_{th}}\right)$$

(2)

where: T_m is the maximum temperature of the peak (K); β = heating rate (K/sec); σ is the trapping cross-section, k_B is the Boltzmann constant, v is the thermal velocity of the charge carriers, N_c is the density of carriers. In the variable heating rate method, the TEES or TSC measurement is repeated at various heating rates and the $\ln(T_{max}^2/\beta)$ quantity plotted as the function of $1/k_B T_{max}$ where T_{max} is the temperature of the current peak maximum. The slope of this curve according to Equation 2 is the thermal ionization energy of the defect level. Using the slope and Y intercept from Equation 2, the capture

cross section can be calculated from Equation 3.

$$\sigma = \exp(-Y_{Intercept})\left(\frac{E_{th}}{N_C v k_B}\right)$$ (3)

A set of variable heating rate TEES and TSC experiments were performed on samples I1,

C and F from which the thermal ionization energy and the trapping cross-section of the

defect levels were extracted.

DISCUSSION

IV measurements

Current-voltage measurements were performed to calculate the bulk resistivity of

samples I1, I2, C and F. Figure 4 is a comparison of the IV characteristics of samples I1,

I2, C and F. Table 2 presents the bulk resistivity results on all the samples thus tested.

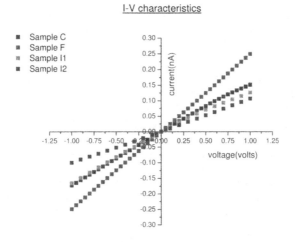

Figure 4. A comparison of current-voltage characteristics for samples I1 (from the

tip region of Ingot I1), I2 (from the mid region of Ingot I2), C (from the tip region of

Ingot B) and F (from the mid region of Ingot B)

Sample	Dopant(s)	Excess %Te	Resistivity (x 10^{10} Ωcm)
Sample I1	In	0	1.89
Sample I2	In	1%	3.7
Sample C	In, Bi	1%	1.88
Sample F	In, Bi	1%	1.26

Table 2. Bulk resistivity of samples I1 (from the tip region of Ingot I1), I2 (from the mid region of Ingot I2), C (from the tip region of ingot B) and F (from the mid region of Ingot B)

While all the samples tested had a bulk resistivity (> 10^{10} Ωcm), the Indium doped 1% excess Te CZT sample I2 had the highest bulk resistivity at 3.7 x 10^{10} Ωcm, and sample F (Bi co-doped In doped CZT sample with 1% excess Te) had comparatively lower bulk resistivity at 1.26 x 10^{10} Ωcm. Also, the electrical resistivity of sample F was lower than that of the sample C (1.88 x 10^{10} Ωcm) but they were high. From the GDMS data for samples C and F as presented in Table 1, except for a variation in the Zn and oxygen concentrations in samples C (3.3% Zn; 65 ppb O) and F (6% Zn; 130 ppb O), the other elemental compositions were similar. It was evident that the bismuth dopant while added to the In-doped CZT crystals (sample C and sample F) did not profoundly affect the bulk resistivity of the crystals.

Detector performance:

A comparison of the spectral response of samples I1 and I2 to gamma radiations from a Co-57 source is presented in Figure 5a. The 14 keV peak and the 122 keV peaks are well defined indicating good electron transport properties in the crystals. Figure 5b presents a comparison of the $\mu\tau_e$ of electrons for samples I1 and I2 respectively, as measured using a Cs^{137} source. Both samples I1 and I2 had a good $\mu\tau_e$ value of 1.56 x 10^{-3} Cm^2/V and 1.5 x 10^{-3} Cm^2/V respectively and were effective radiation detectors.

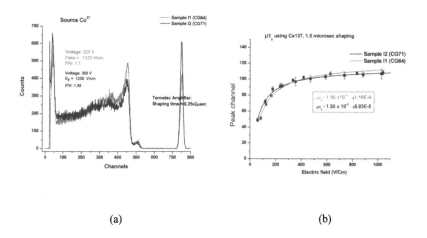

(a)   (b)

Figure 5 a. Spectral response to gamma radiations from Co^{57} source of sample I1 and I2; b. A comparison of $\mu\tau_e$ of samples I1 and I2 using Cs^{137} source.

A comparison of the spectral response of samples C and F (from Bi co-doped Indium doped CZT ingot B) to gamma radiations from a Co-57 source is presented in Figure 6. Though the crystals exhibited high bulk resistivity and an ability to hold high electric fields, they showed a poor spectral resolution for the Co-57 photo peaks. Also, samples C and F gave a poor spectral response for the Cs^{137} source. Hence, $\mu\tau_e$ measurements could not be performed.

146

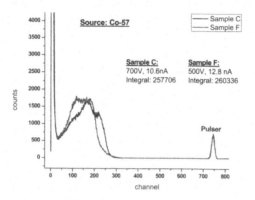

Figure 6. Spectral response to Co-57 source for sample C and sample F

The poor spectral response of the Bi co-doped Indium doped CZT crystals to Co-57 and Cs-137 sources indicated severe trapping of charge carriers which rendered them unusable as radiation detectors.

Spectroscopy studies (TEES/TSC)

Though all the samples tested exhibited a high bulk resistivity ($>10^{10}$ Ω.Cm), the Bismuth co-doped Indium doped CZT samples were poor radiation detectors compared to the Indium doped CZT crystals which exhibited excellent radiation detection capabilities. This indicated the presence of deep level defect (or defects) in the band gap of the Bi co-doped Indium doped CZT samples that could act as trapping centers for charge carriers. TEES and TSC experiments were performed on samples I1, I2, C and F to study the band gap defect levels in the crystals. Figure 7 presents the results from TEES experiments on samples I1, I2, C and F performed at a heating rate of 0.072 K/sec. The TEES data for the Indium doped samples I1 and I2 revealed two prominent levels labeled I1 and I2. The TEES data for samples C and F (from the tip and the mid of the ingot B) have revealed many prominent levels as labeled in Figure 7. Also, from Figure 7, a positive current at

147

room temperature (and above it) indicated a p-type conductivity in all the samples. Holes were the dominant charge carriers in both the Indium doped, as well as, the bismuth co-doped Indium doped samples.

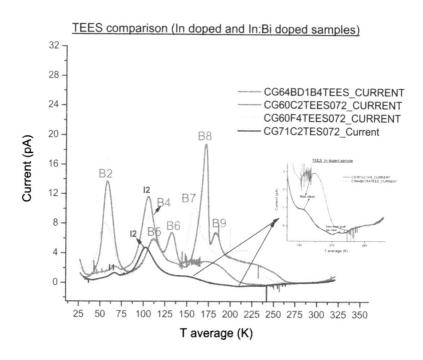

Figure 7. TEES results for samples I1, I2, C and F; performed at a heating rate of 0.072 K/sec.

Figure 8 presents the thermally stimulated current (TSC) results on samples I1, I2, C and F performed at a heating rate of 0.072 K/sec. The TSC data from Indium doped samples have revealed the following levels labeled I1, I2, I3, and I4. While the current peaks I1 and I2 were prominent, current peaks I3 and I4 were poorly resolved.

The TSC data from the Bi co-doped Indium doped samples C and F have revealed several prominent peaks labeled B1 through B10 in Figure 8.

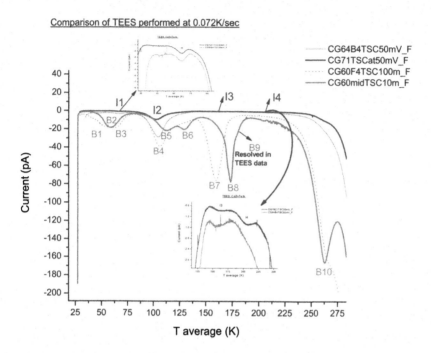

Figure 8: TSC plots for samples I1 and I2, both performed at a heating rate of 0.072 K/sec and 50mV TSC voltage; Sample C (performed at 0.072 K/sec and 10 mV TSC voltage); and sample F (performed at 0.072 K/sec and 100 mV TSC voltage).

Also, variable heating rate TSC experiments were performed on the Indium doped sample I1, and Bi co-doped samples C and F to calculate the thermal ionization energy and the trap cross-section of the various defect levels. Figure 9 a presents a variable

149

heating rate TSC plot obtained from sample C for the defect level labeled B8 (in Figure 8). The temperature (T_{max}) corresponding to each current peak maximum and the respective heating rate were recorded. Using this data an Arrhenius plot was created for the defect level observed. This is presented in Figure 9b. From the slope of this plot, and using Eqn. 2, the thermal ionization energy and the capture cross-section of the defect level B8 was calculated to be (0.39 ± 0.02)eV and (2.54 ± 5) x 10^{-16} cm^2 respectively.

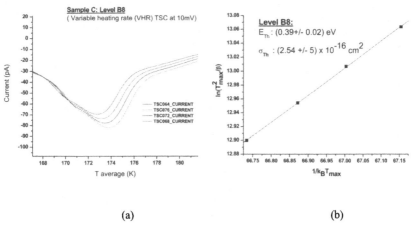

(a)   (b)

Figure 9 a: variable heating rate TSC plot for level B8 in sample C (performed at 0.072 K/sec and 10mV TSC voltage). b: The corresponding Arrhenius plot from which the thermal ionization energy (E_{Th}) and the trap cross-section (σ_{Th}) of the defect level B8 was calculated.

Table 3 presents the thermal ionization energy and the capture cross-section of levels I1, I2, and I3 as calculated from the variable heating rate TSC plots from the Indium doped sample I1. Though peaks I1 through I4 were observed in the TSC plot for sample I2, variable heating rate TSC experiments were not performed at this time.

Levels → Samples ↓	I1	I2	I3	I4
Sample I1 **Energy** **Trap Cross-section**	**Peak:67K** (0.1±0.02) eV (6.4±5) x 10^{-19} cm^2	**Peak:104K** (1.6±0.02) eV (3.3±5) x 10^{-19} cm^2	**Peak at 160K** Peak poorly resolved.	Not Observed
Sample I2	Peak observed. but VHR TSC not performed	Peak observed. but VHR TSC not performed	Peak poorly resolved. VHR TSC not performed	Peak poorly resolved. VHR TSC not performed

Table 3. Thermal ionization energy and trap cross-section of prominent levels in Indium doped samples I1 and I2.

Levels → Samples ↓	B1	B2	B3	B4	B5	B6	B7	B8	B9	B10
Sample C	No	**Peak:58 K** 0.08±0.02 eV (2.0±5) x 10^{-18} cm^2	No	No	**Peak:112 K** 0.19±0.02 eV (8.2±5)x 10^{-19} cm^2	**Peak: 130K** 0.27±0.02 eV (1.4±5)x 10^{-16} cm^2	No	**Peak:172 K** 0.39±0.02 eV (2.54±5)x 10^{-16} cm^2	**Peak:183 K** 0.44±0.02 eV (1.5±5)x 10^{-15} cm^2	**Peak:262 K** 0.82±0.02 eV (2.59±5)x 10^{-12} cm^2
Sample F	**Peak:43 K** Peak poorly resolved.	No	**Peak:62 K** Peak poorly resolved.	**Peak:106 K** 0.19±0.02 eV (8.2±5) x 10^{-19} cm^2	No	No	**Peak:158 K** 0.33±0.02 eV (6.2±5)x 10^{-17} cm^2	No	No	Peak poorly resolved.

Table 4. Thermal ionization energy and trap cross-section of prominent levels in Bi co-doped Indium doped samples C and F.

Table 4 presents the thermal ionization energy and the trap cross-section of levels B1 through B10 as calculated from the variable heating rate TSC plots from the Bi co-doped Indium doped samples C and F (from the tip and mid of ingot B respectively). Given the margin of error in the calculation for the thermal ionization energy and the trap cross-section of a defect level, a comparison of the shallow energy levels from Table 3 and Table 4 did not help distinguish if a particular shallow defect level was associated with the Indium doped sample or the Bi co-doped Indium doped sample. Nevertheless, in the absence of deep level traps in the Indium doped samples I1 and I2, defect levels B6 (with a thermal ionization of 0.27 ±0.02 eV and a trap cross-section of $(1.45 \pm 5) \times 10^{-19}$ cm^2), B7 (with a thermal ionization of (0.33 ± 0.02) eV and a trap cross-section of $(6.2 \pm 5) \times 10^{-17}$ cm^2), B8 (with a thermal ionization of (0.39 ± 0.02) eV and a trap cross-section of $(2.5 \pm 5) \times 10^{-16}$ cm^2), B9 (with thermal ionization energy of (0.44 ± 0.02) eV and trap cross-section of $(1.5 \pm 5) \times 10^{-15})$ and B10 (with a thermal ionization of (0.82 ± 0.02) eV and a trap cross-section of $(2.59 \pm 5) \times 10^{-12}$ cm^2) were associated with bismuth doping in Indium doped CZT crystals.

While TEES/TSC techniques were employed to study the bismuth induced defect levels in the Bi co-doped Indium doped samples, characterization of the nature or determination of the type of defect (or defect complex) associated with each defect level was beyond the scope of this research. The unipolar emission in the TEES spectrum of the samples makes it difficult to determine if the observed trap level is emitting holes or electrons. Considering the deep level B10 with a thermal ionization energy of (0.82 ± 0.02) eV and with a large trap cross-section (the order of 10^{-12} cm^2), this is a strong trapping center for charge carriers. It could be a major contributing defect level responsible for the deterioration in the effective detector performance of the Bi co-doped Indium doped crystal.

CONCLUSIONS

The performance of Indium doped CZT samples and bismuth co-doped Indium doped CZT samples as room temperature radiation detectors were evaluated. While the Bi co-doped Indium doped crystals exhibited a high bulk resistivity (> 10^{10} Ωcm), they were poor radiation detectors. The Indium doped crystals tested were very good radiation detectors with an average bulk resistivity of 2.79 x 10^{10} Ωcm and $\mu\tau_e$ of 1.52 x 10^{-3} cm^2/V respectively. TEES/TSC techniques were used to study the defect levels in the Indium doped and Bi co-doped indium doped samples. By comparing the thermal ionization energy and trap cross-section of various prominent peaks observed in all the samples, the levels that were associated only with Bismuth doping in Indium doped samples were identified. Among these defect levels, the deep level B10 with a thermal ionization energy of (0.82 ±0.02) eV and a trap cross-section of (2.59 ±5) x 10^{-12} cm^2 was identified as a strong trapping center for charge carriers. This level is suggested to be the major cause for the poor performance of the Bi co-doped Indium doped crystals as room temperature radiation detectors.

ACKNOWLEDGEMENTS

The authors thank the US Department of Energy, NA-22, Contracts DE-FG52-06/27497/A000 and DE-FG52-08NA28769 for their financial support of this research.

REFERENCES

1. Cs. Selez, Y.Y. Shan, K.G. Lynn, A.R. Moodenbough, and E.E. Eissler, Phys. Rev. B 53, 6495 (1997).
2. P. Emanuelson, P. Omling, B.K. Meyer, M. Wienecke, and M. Schenk, Phys. Rev. B 47, 15578 (1993).
3. N. Krsmanovic, K.G. Lynn, M.H. Weber, R. Tjossem, Th. Gessmann, Phy. Rev. B 62, (24) 16279 (2000).
4. R. Soundararajan, K.G. Lynn, S. Awadallah, C, Szeles, Su-Huai Wei, J. Elect. Mat. 35, (6) 1333 (2006).
5. Su-Huai Wei, and S.B. Zhang, Phy. Stat. Sol. B 229 (1), (2002).
6. Minoru Funaki, Yukio Ando, Ryuji Jinnai, Akira Tachibana, and Ryoichi Ohno, http://www.acrorad.co.jp/pdf/Development_of_CdTe_detectors.pdf
7. S. Terterian, M. Chu, D. Ting, L.C. Wu, C.C. Wang, M. Szawlowski, G. Vissor, and P.N. Luke, J. Elec. Mat. 32, p796 (2003).
8. E. Saucedo, L. Fornaro, N.V. Sochinskii, A. Cuna, V. Corregidor, D. Granados, E. Dieguez, IEEE Trans. Nuc. Sci 51, (6) (2004).
9. E. Saucedo, O. Martinez, C.M. Ruiz, O. Vigil-Galan, I. Benito, L. Fornaro, N.V. Sochinskii, E. Dieguez, J. Cry. Gro. 291, 416 (2006).
10. Jong Hee Suh, Sin Hang Cho, Jae Ho Won, Jin Ki Hong, Sun Ung Kim, J. Kor. Phy. Soc. 49 (2006).

11. M. Bliss, D.C. Gerlach, J.B. Cliff, M.B. Toloczko, D.S. Barnett, G. Ciampi, K.A. Jones, K.G. Lynn, Nuc. Inst. Met. Phy. Res. A **579** (2007).
12. O. Vigil-Galan, M. Brown, C.M. Ruiz, M.A. Vidal-Borbolla, R. Ramirez-Bon, E. Sanchez-Meza, M. Tufino-Velazquez, M. Estela Calixto, A.D. Compaan, G. Contreras-Puente, Thin Sold Films **516** (2008).
13. V. Babentsov, J. Franc, P. Hoschl, M. Fiederle, K.W. Benz, N.V. Sochinskii, E. Dieguez, and R.B. James, Cryst. Res. Technol., 1-5 (2009)
14. Mao-Hua Du "Bismuth-induced deep levels and carrier compensation in CdTe" Phy. Rev. B **78**, (1995).
15. E. Saucedo, J. Franc, H. Elhadidy, P. Horodysky, C.M. Ruiz, V. Bermudez and N.V. Sochinskii, J. Appl. Phy. **103**, (2008).
16. R. H. Bube, "Photoconductivity in Solids", p. 242 (New York: Wiley, 1960).

Mater. Res. Soc. Symp. Proc. Vol. 1164 © 2009 Materials Research Society   1164-L10-04

Recent advances in THM CZT for Nuclear Radiation Detection

J. MacKenzie[1], H. Chen[1], S. A. Awadalla[1], P. Marthandam[1], R. Redden[1], G. Bindley[1], Z. He[2], D. R. Black[3], M. Duff[4], M. Amman[5], J. S. Lee[5], P. N. Luke[5], M Groza[6], and A. Burger[6]

[1]Redlen Technologies, Sidney, BC V8L 5Y8 Canada
[2]University of Michigan, Ann Arbor, MI 48109 USA
[3]National Institute of Standards and Technology, Gaithersburg, MD 20899 USA
[4]Savannah River National Laboratory, Aiken, SC 29808 USA
[5]Lawrence Berkeley National Laboratory, Livermore, CA 94720 USA
[6]Fisk University, Nashville, TN 37208 USA

Abstract:

Greater than 500 cm^3 single crystal CdZnTe (CZT) has been grown using the travelling heater method providing thick (> 5 mm) detectors required for high energy gamma ray detection. Detectors greater than 5 mm thickness have achieved superior energy resolution or < 1 %FWHM for 662 keV gammas from a ^{137}Cs source while maintaining this level of resolution for 1.17 and 1.33 MeV gammas from a^{60}Co source. Standard, 5 mm thickness detectors also show increases in mobility-lifetime products for both electrons and holes in crystals grown over 3 years. The improvement in these measurable is attributable to improvements in quality of the grown CZT and post growth heat treatments.

Introduction:

Currently, nuclear radiation detection and application still mainly employ scintillation detectors and solid state high purity Germanium (Ge) based detectors. These detectors, however, suffer from several drawbacks including the need for cryogenic cooling to reduce thermally generated leakage currents in Ge. The cooling requirement for Ge and photomultiplier tubes required for scintillator makes these detection systems bulky; requiring significant infrastructure costs for operation. Large scale deployment of gamma ray spectroscopy applied to areas of nuclear medical imaging, homeland security and astrophysics is hindered by the large size and infrastructure required for these detection systems.

Cadmium Zinc Telluride (CZT) has been identified as a promising material for room temperature gamma and X-ray spectroscopy based on having a wide band-gap suitable for detecting such high energy photons. High resistivity CZT has been shown to have good electron collection properties and long term stability without suffering temperature polarization effects associated with room temperature operation [1]. Melt grown High and low pressure Bridgman (HPB, LPB respectively) are known to produce CZT suitable for high resolution gamma and X-ray spectroscopy and much effort has been devoted to commercialization of CZT using these growth techniques [2]. Unfortunately, the large cost and low yield associated with melt growth CZT techniques has impeded mass production.

CZT grown using the solution growth, travelling heater method (THM) has been shown to produce CZT of equal or superior electrical properties suitable for gamma and X-ray spectroscopy [3]. Traditional prejudice held that the THM process was inappropriate for commercial production due to the slow growth rates and historically small sizes grown. However, these obstacles have been overcome recently and single crystal CZT volumes of >260 cm^3 are now routinely produced. This increase in volume has paralleled an improvement in the electrical properties of THM grown CZT and desire for increased thickness detectors suitable for higher energy gamma ray detection. This paper highlights the advancements and challenges associated with THM grown CZT single crystal volume increases, defect characterization and improvements in electrical properties.

Advancements in single crystal volume:

To date, only Redlen Technologies grow CZT using THM. 50 and 75 mm diameter crystals having single crystal volumes of ~140 and 260 cm^3 are routinely produced (Figure 1).

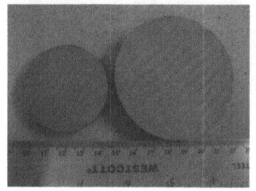

Figure 1: THM grown, 50 (left) and 75 mm diameter 100% single crystal CZT wafers.

Demand for increased thickness CZT (Figure 2) for higher energy gamma detection requires increased CZT volume. Increased volumes can be accommodated by increasing boule diameter or height (or both). Most recently, 100 mm diameter single crystal CZT having a volume in excess of 500 cm^3 has been grown using THM. Overall, 75 – 100 mm length boules are grown and subsequently wafered to the desired thickness suitable to produce the required thickness after post growth heat treatments. The advancements in ingot singularity, diameter and height have facilitated production of oriented, 20 x 20 x 15 mm CZT detectors. Heat treatments for wafers of >5 mm thickness have yet to be optimized but have routinely produced detectors having 1-2% FWHM, 662 keV, ^{137}Cs gamma spectra.

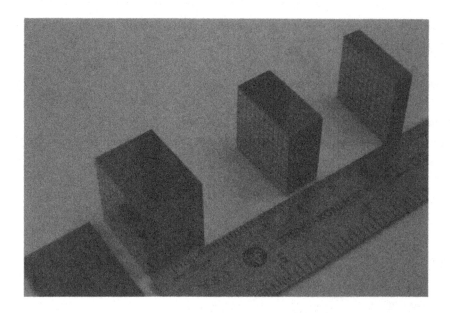

Figure 2: from left to right, 20 x 20 x 15, 10 and 5 mm thick CZT.

While macro scale defects such as polycrystallinity and twinning have largely been eliminated through development of improved crystallinity seeds and improvements in the growth conditions, the major remaining current challenges include: 1) inhomogeniety in electrical transport properties, 2) micro-scale Tellurium decorated linear (Figure 3) and planar defects and 3) development of other extended defects including non-uniform strain levels (Figure 4A) and mosaic structures (Figure 4B). Mosaic structures represent a formation of a dislocation net within the crystal. Te decorated mosaic and linear structures are believed to form during periods of constitutional super cooling and may reflect transient changes in convective flow in the solvent during growth.

While these challenges are significant, they are resolvable and are gating factors incurred in expanding the grown crystal volume and ultimately launching a commercial scale growth process.

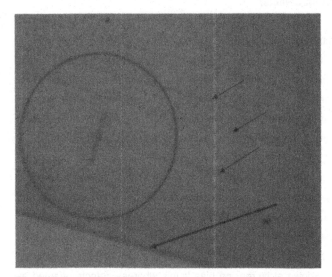

Figure 3: Transmitted IR image (unpolarized) of Te decorated linear defects in THM grown CZT. 1 mm scale bar is located inside 3 mm diameter circle. Short arrows point to some of the Te lineations while the long arrow shows the planar Te decorated feature.

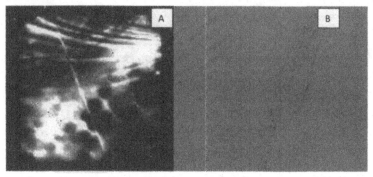

Figure 4: Cross polarized transmitted IR image of a 10 x 10 mm^2 tile showing non-uniform distribution of strain (left) and mosaic structure viewed using unpolarized transmitted IR (right). FOV for figure 4B is 0.12 mm^2.

Electrical effects of current defects:

The current defects noted above have variable effects on detector performance. Specifically, it is not only the presence of one or more of these defects but their location and extent are also important. It should be noted that the remaining defects noted appear to have a stronger effect on 10 and 15 mm thickness. Redlen's standard product is 20 x20 x5 mm^3 CZT having 8 x 8 pixels, with pixel size of 2x2mm^2 and 2.46 mm pitch, and an Au contact. Figure 5 shows the typical gamma spectra obtained using such a detector.

For 10 and 15 mm thickness detectors, linear and planar Te decorated defects physically trap electrons, deteriorating final device performance. If these features are located near the cathode, excessive high cathode noise has been observed, especially if the defect intersects the actual cathode surface. The orientation of the defect is also important. Linear and planar defects oriented at high angles relative to the bias direction may promote device breakdown at increasing bias levels whereas defects oriented at shallower angles may cause charge loss and local side drift.

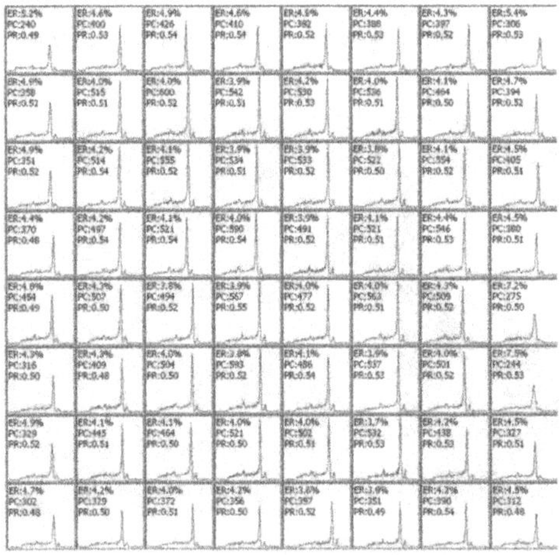

Figure 5: Typical 20 x 20 x 5 mm3 pixellated THM CZT detector. Room temperature [57]Co, 122 keV response (700V bias, no signal processing correction)

The presence of mosaic and other 3-D extended structures as well as inherited strain within 10 and 15 mm thickness CZT detectors cause local deteriorations of electron charge collection and distortion of the electrical field all of which act to degrade detector performance.

Further to the remaining intrinsic defects sourced by the growth processes are extrinsic surface features which also act to degrade device performance. X-ray topography shows that even while the polished surfaces may be free of any visible scratches, a strain field may remain which may perform as a parallel resistance path leading to the reduction of the total device resistance and increase the dark leakage current.

Material quality improvement:

While increased thickness detectors are coming on-line, it is also important not to lose sight of performance standards derived from standard 20 x 20 x 5 mm^3 detectors. Figure 6 is a plot of Mu-tau-e (blue) and Mu-tau-h (green) over three years of production. The figure shows that electron AND hole mobility-lifetime products have both increased with increasing production and optimization of 20 x 20 x 5 mm^3 detectors. It may appear contradictory that the mobility-lifetime products of both electrons and holes increase similarly but the two need not be opposites of one another as global improvements in crystallinity and reduction of intrinsic defects act to improve overall transport properties which affect both.

In addition to improvements in Mu-tau-e and Mu-tau-h, THM grown CZT has also improved with respect to gamma resolution. THM grown CZT approaches theoretical limits of energy resolution [4] with no downstream signal processing. Figure 7 shows that THM grown CZT can resolve 662 keV gammas from ^{137}Cs source AND 1.17 / 1.33 MeV gammas from a ^{60}Co source to better than 1% FWHM at room temperature. To obtain such sharp energy resolutions previously required much thicker detectors as well as additional signal processing.

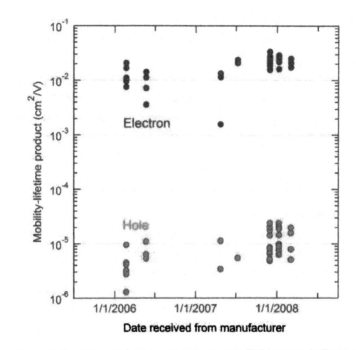

Figure 6: Evolution of Mu-tau-e and Mu-tau-h in THM grown CZT with date of manufacture.

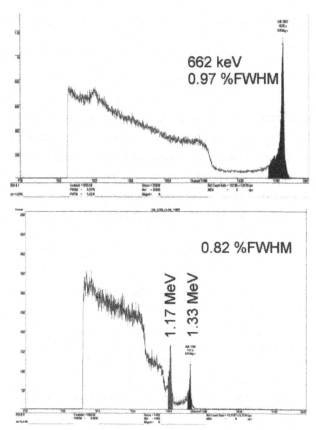

662 keV
0.97 %FWHM

0.82 %FWHM

1.17 MeV
1.33 MeV

Figure 7: Energy response of 3x3x6 mm³using MCA :a) ¹³⁷Cs (662 keV) with FWHM of 0.97% at 1200V for 120 s real time and 0.5 μs shaping time and b) ⁶⁰Co (1.17 and 1.33 MeV) with 0.82% for 1.33 MeV at 1500V, 300 s real time and 0.5 μs shaping time. Data was taken room temperature without any additional signal correction.

Finally, THM grown CZT has been evaluated using alpha particles from ²⁴¹Am. The resultsof these tests show that THM grown CZT has stable Mu-tau-e over a temperature range of 0-30℃ (10.67 and 11.13 %FWHM at 0℃ at 30℃ respectively). Improvements in material processing and fabrication have also lead to reductions in leakage currents, increasing inter pixel resistance and improved sensitivity (peak counts).

Conclusions:

THM grown CZT has expanded to routinely produce single crystal volumes in excess of 260 cm^3. Increases in singularity and volume have facilitated fabrication of thicker detectors for high energy gamma energy resolution useful for homeland security and other applications. Perhaps more importantly, advances in crystal growth, post growth heat treatment and fabrication have all combined to provide THM CZT with improved electrical transport properties and energy resolution at room temperatures, a cardinal goal of nuclear radiation detection.

References:

[1] C. Szeles and M.C. Driver (1998) Growth Properties of semi-insulating CdZnte for radiation detector applications, Proceedings of SPIE, Vol. 3446, 2-9.

[2] H. Chen, S.A. Awadalla, J MacKenzie, R. Redden, G. Bindley, A.E. Bolotnikov, G.S. Camarda, G. Carini, R.B. James (2007): Characterization of Travelling Heater Method (THM) Grown $Cd_{0.9}Zn_{0.1}Te$ Crystals. IEEE Transactions on Nuclear Science, V. 54, No. 4, 811-816.

[3] M. Amman, J.S. Lee, P.N. Luke, H. Chen, S.A. Awadalla, R. Redden and G. Bindley (2009) Submitted.

[4] P. Dorenbos (2002) Light output and energy resolution of Ce3+-doped scintillators. Nuclear Instruments and Methods in Physics Research, A486, 208-213.

Mater. Res. Soc. Symp. Proc. Vol. 1164 © 2009 Materials Research Society   1164-L10-07

Opto-electrical characterization and X-ray Mapping of large-volume cadmium zinc

telluride radiation detectors

G. Yang, A. E. Bolotnikov, G. S. Camarda, Y. Cui, A. Hossain, H. W. Yao, K. Kim and R. B. James

Brookhaven National Laboratory, Upton, NY 11973, U.S.A.

ABSTRACT

Large-volume cadmium zinc telluride (CZT) radiation detectors would greatly improve radiation detection capabilities and, therefore, attract extensive scientific and commercial interests. CZT crystals with volumes as large as hundreds of centimeters can be achieved today due to improvements in the crystal growth technology. However, the poor performance of large-volume CZT detectors is still a challenging problem affecting the commercialization of CZT detectors and imaging arrays. We have employed Pockels effect measurements and synchrotron X-ray mapping techniques to investigate the performance-limiting factors for large-volume CZT detectors. Experimental results with the above characterization methods reveal the non-uniform distribution of internal electric field of large-volume CZT detectors, which help us to better understand the responsible mechanism for the insufficient carrier collection in large-volume CZT detectors.

INTRODUCTION

Cadmium zinc telluride (CZT) has been considered a promising material for room-temperature nuclear radiation detection, since it provides high detection-efficiency and good energy-resolution without a complicated cooling system [1-3]. Since the first practical CZT gamma-ray detector reported in 1992 [2], there have been many advances in the performance of the devices. However, most of previous investigations were focused on small-volume CZT detectors because of the limited availability of large-volume CZT single crystals. Actually large-volume CZT detectors are always desired because they substantially improve the detection-efficiency and reduce the measurement time, which are especially important in hand-held gamma-ray spectrometers, medical imaging systems and astrophysics experiments. Unfortunately, even though recent developments of crystal growth techniques provide a better availability of large-volume CZT crystals, it is still difficult to fabricate large-volume CZT detectors with high carrier collection efficiency. An important parameter that substantially affects the carrier collection efficiency and, therefore, the performance of CZT detectors, is the distribution of the internal electrical field; a uniform distribution is always desirable. However, to our knowledge, until now there are few reports to investigate the electric field distribution of large-volume CZT detectors. In this work, we employed a Pockels effect (PE) measurement system and synchrotron X-ray mapping technique to investigate the electric field distribution of a large-volume CZT detector. To avoid the complexity introduced by different contact configurations, our research is focused on Au–CZT planar detectors, because the planar configuration of contacts is the basic one used for commercial nuclear radiation detectors. The understanding of the distribution of internal electric field of planar detectors is also meaningful to other detectors with different contact configurations.

EXPERIMENT

A large-volume CZT crystal, 10 x10 x 10 mm^3, was investigated in this work, which was grown by the modified low-pressure Bridgman method. Following mechanical polishing by 0.05-μm particle-size alumina suspension and chemical polishing by 2% bromine-methanol solution, Au was deposited on the top surface and the bottom surface of the CZT sample. In this way a planar CZT detector was fabricated. To describe the distribution of internal electric field more clearly, we set up an x-y-z coordinates relative to the position of the CZT detector (shown in figure 1). Next we employed a PE measurement system, illustrated in figure 2, to reflect the corresponding lateral distribution, i.e. z-direction distribution, of the internal electric field. In this system, a collimated Xe lamp with a 950-nm infra-red (IR) filter illuminated the entire lateral surface of the CZT detector. Two linear polarizers separately acted as the polarizer and the analyzer. The transmitted light was focused on a charge coupled device (CCD) camera controlled by SynerJY software and then generated the PE images.

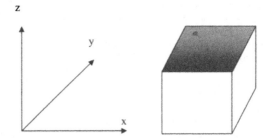

Figure 1. x-y-z coordinates for the description of internal electric field of the large-volume CZT detector

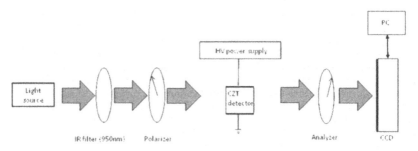

Figure 2. Schematic diagram of the Pockels effect measurement system

During the X-ray mapping measurements at Brookhaven's National Synchrotron Light Source (NSLS), an X-ray beam from a synchrotron radiation ring, with a spot size of $10 \times 10 \ \mu m^2$, was used to irradiate the CZT detector. For each position of the beam, the corresponding energy spectra and its associated information (i.e., pulse height, photopeak position, and FWHM) were collected. By raster-scanning the CZT detector in the x- and y-directions and plotting the detector's photopeak positions over its entire area, we acquired X-ray response maps. Because the response is related to charge collection, and also strongly affected by the internal electric field, such measurements yield information on the field's two-dimensional distribution (x-y direction distribution).

RESULTS AND DISCUSSIONS

In the PE measurement, the top planar electrode and the bottom planar electrode were respectively treated as the anode and the cathode of CZT detector. Before applying the bias voltage, the PE image is completely dark, indicating there is no leakage light disturbance during this measurement. Figure 3 shows the PE image of the CZT detector with a bias voltage of +2500V. According to Guenther [4], the intensity of transmitted light passing through the crossed polarizer and analyzer can be described by

$$I = I_0 \sin^2 (\frac{\pi \, n_0^3 rd}{\lambda} E), \tag{1}$$

where I_0 is the maximum light intensity passing through uncrossed polarizers, n_0 is the field-free refractive index, r is the linear electro-optic coefficient for CZT, d is the light path length through CZT crystal, λ is the free space wavelength of the illuminated IR light, and E is the mean electric-field along the optical path.

Since $\frac{\pi \, n_0^3 rd}{\lambda} E \ll 1 [5]$, we have

$$I \approx I_0 (\frac{\pi \, n_0^3 rd}{\lambda} E)^2 \propto E^2 \tag{2}$$

and

$$E \propto \sqrt{I} \tag{3}$$

As a result, the intensity distribution of the PE image indicates the lateral distribution of the internal electric field. From figure 3, one can see there are obvious changes of the internal electric field when a bias voltage of +2500V is applied. The amount of photon counts indicates that the intensity of the internal electric field was enhanced with increasing bias voltage. More importantly, we found the distribution of the internal electrical field is not uniform; the internal electric field firstly increases laterally from the anode towards the cathode until it arrives at the maximum value, which is approximately located on the one third of the full height, then generally decreases towards the cathode. This behavior is somewhat similar with the internal electric field in CdTe detector equipped with In and Pt contacts [6]. To further achieve the details of internal electric field distribution, we also plotted the intensity distribution from the left side to the right side (along C-D line in figure 3) and observed another interesting phenomenon. Compared with the central section, the internal electric field is much stronger near the left edge and right edge. In other words, the strongest electric field was 'defocused' from the center toward both edges when the bias voltage of +2500V was applied.

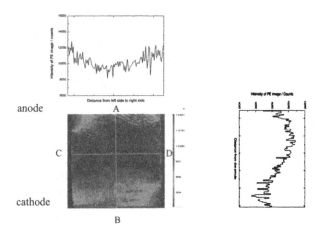

anode

C   D

cathode

B

Figure 3. PE image of the large-volume CZT detector under the bias voltage of +2500V

To address this issue, we further employed the micron-scale X-ray mapping system developed at NSLS to investigate the two-dimensional distribution of the internal electric field (x-y directions). Figure 4 shows the X-ray response map of the large-volume CZT detector, where a positive bias voltage of 1000V was applied on both electrodes and a synchrotron beam of 26.9-keV X-rays was incident on the sample. Obviously, the central section of the X-ray map is uniformly white, while the surrounding edges are grey, even dark. It means the internal electric field near the edges is not uniform, and a poorer carrier collection was incurred in these edge regions. This observation is consistent with the above 'defocusing phenomenon" revealed by Pockels effect measurement. We consider it is due to the fact that the electrostatic potential on the detector's side surfaces decreases faster than the electrostatic potential along the detector's axis, which result in the field lines strongly bent towards the side surfaces. Clearly, in this case, the carriers generated by incident particles near the edges are driven to the side surface, where they become trapped. As a result, this "defocusing" electric field causes an undesirable charge loss near the device's edges in large-volume CZT detectors.

Figure 4. X-ray response map of the large-volume CZT detector

CONCLUSIONS

Pockels effect measurements and a synchrotron X-ray mapping technique were used to investigate the internal electric field of large-volume CZT radiation detectors. The experimental results indicate that the internal electric field in the large-volume CZT detector is non-uniform, and a defocusing effect of the field exists. For achieving better carrier collection efficiency, it is necessary to improve sample surface processing techniques and/or explore new device designs to correct the non-uniformity of the internal electric field of large-volume CZT detectors.

ACKNOWLEDGMENTS

This work was supported by US Department of Energy, Office of Nonproliferation Research and Development, NA-22. The manuscript has been authored by Brookhaven Science Associates, LLC under Contract No. DE-AC02-98CH1-886 with the U.S. Department of Energy. The United States Government retains, and the publisher, by accepting the article for publication, acknowledges, a worldwide license to publish or reproduce the published form of this manuscript, or allow others to do so, for the United States Government purposes.

REFERENCES

1. R. B. James, T. E. Schlesinger, J. C. Lund, and M. Schieber, Semiconductors for Room Temperature Nuclear Detector Applications, Academic Press, New York, 1995, Vol. 43, p. 334.
2. F. P. Doty, J. F. Butler, J. F. Schetzina and K. A. Bowers, J. Vac. Sci. Technol. B10, 1418 (1992).
3. G. Yang, A. E. Bolotnikov, Y. Cui, G. S. Camarda, A. Hossain, and R. B. James, J. Cryst. Growth 311, 99 (2008).

4. R. Guenther, Modern Optics (New York: John Wiley & Sons, Inc., 1990), pp. 569-590.
5. H. W. Yao, R. J. Anderson, and R. B. James, Proc. SPIE 3115, 62 (1997).
6. A. Cola, I. Farella, A. M. Mancini, and A. Donati, "Electric Field Properties of CdTe Nuclear Detectors", IEEE Trans. Nucl. Sci. 54, 868 (2007).

Scintillator II

Mater. Res. Soc. Symp. Proc. Vol. 1164 © 2009 Materials Research Society   1164-L11-01

Structure and Luminescence of Europium-Doped Gadolinia-Based Core/Multi-Shell Scintillation Nanoparticles

Teng-Kuan Tseng, Jihun Choi and Paul H. Holloway
Department of Materials Science and Engineering
University of Florida, Gainesville, FL 32611-6400, USA

ABSTRACT

Scintillating nanoparticles with a SiO_2 core and a Gd_2O_3 shell doped with Eu^{3+} were synthesized with a sol-gel process. Based on transmission electron microscopy (TEM) data, a ~13 nm Gd_2O_3 shell was successfully coated onto ~220 nm mono-dispersed SiO_2 nanocores. Eu^{3+} ions at concentrations of nominally 5 at% exhibited photoluminescent (PL) emission from the SiO_2/Gd_2O_3 nanoparticles after being calcined at 800 ^0C for 2 h. The SiO_2 remained amorphous after calcining, while the Gd_2O_3 crystallized to a cubic structure. The PL emission was from the $^5D_0$-$^7F_2$ transitions of Eu^{3+} at 609 and 622 nm. Photoluminescence excitation (PLE) data showed that emission from Eu^{3+} could result from direct excitation, but was dominated by the oxygen to europium charge-transfer band (CTB) between 250 and 280 nm for Eu^{3+} doped in Gd_2O_3. The quantum yield (QY) from thin films drop cast from a mixture of 20 mg of calcined nanoparticles in 500 μL of polymethylmethacrylate (PMMA) and excited in the CTB was 20% for $SiO_2/Gd_2O_3:Eu^{3+}$ core/shell scintillation nanoparticles. Finally, the above core/shell nanoparticles were passivated with a shell of SiO_2 to create e.g. $SiO_2/Gd_2O_3:Eu^{3+}/SiO_2$ nanoparticles. The QYs for this nanostructure were lower than unpassivated nanoparticles which was attributed to a weak CTB for the amorphous SiO_2 shell and a higher density of interface quenching sites.

INTRODUCTION

Luminescent rare-earth nanocrystals and colloidal quantum dots incorporated with dopants have aroused great interests due to their applications in various fields, i.e. display and solid-state lightening, scintillator detector, and biologic diagnosis [1-4]. In recent years, core/shell nanomaterials also have been studied extensively because of their great versatility in applications such as being tuned for high catalytic activity, better stability in various environments, and controllable surface modification properties. There are a variety of examples of core/shell nanostructures possessing merits mentioned above, e.g. Au/SiO_2, $CdSe/SiO_2$, CdS/ZnS, and polypyrole/Fe_2O_3 [5-7]. For applications as scintillation detectors, the materials should possess the properties of being luminescent with short decay time, high atomic number and density, low afterglow, large light output and low hygroscopicity. Generally, the scintillation materials are made in the form of single crystal by elaborate single crystal growth methods, i.e. Czochralski or Bridgman growth methods. In this work, a sol-gel precipitation method will be employed to prepare $SiO_2/Gd_2O_3:Eu^{3+}$ and $SiO_2/Gd_2O_3:Eu^{3+}/SiO_2$ core/multi-shell scintillation nanoparticles. Mono-dispersed spherical SiO_2 nanoparticles were first prepared to serve as the template and Gd_2O_3 shell was deposited on the surfaces of SiO_2. Europium doped Gd_2O_3 shells were prepared and studied using luminescent properties correlated with the effect of charge transfer from oxygen to europium. The correlations obtained are of prime importance to realize an efficient and stable utilization of nanocomposite scintillation materials.

EXPERIMENTAL DETAILS

Materials synthesis

Silica nanoparticles were fabricated via the Stober method [8], in which 10 mL of concentrated ammonium hydroxide (NH_4OH) was mixed with 200 mL of ethanol and 40 mL of deionized (DI) water and stirred thoroughly. 20 mL of tetraethoxysilane (TEOS) solution was added and stirred vigorously 40 min. No visible change occurs at first but after a few minutes, the solution became hazy and then opaque white as the silica particles grow large enough to scatter light. Mono-dispersed 220~230 nm spherical silica nanoparticles were obtained and used as cores for further coating.

Luminescent gadolinium oxide shell was coated onto the silica cores using a sol-gel method. A concentration of 0.4g/100mL of silica nanoparticles was prepared and sonicated to have a homogeneous suspension, followed by additions of 0.04 mol/L gadolinium nitrates and 0.008 mol/L europium nitrates mixed with 1 mol/L urea to yield a transparent solution that was stirred at 82~85 ℃. After being vigorous stirring for 2 h, white precipitates were collected by washing and centrifuging with DI water three times. The products were oven-dried at 80 ℃ for 2 h, followed by an 800 ℃ calcination in air for 2 h.

Materials characterization

The phase structures of as-prepared and calcined samples were characterized by X-ray diffraction (XRD) (Philips APD 3720) with Cu K_α radiation source (λ=0.5418 nm). The XRD pattern was collected from powder samples in the step scan (0.02˚) mode with a small grazing incident angle over a 2θ scan range of 20-70˚. The morphology and particle size were determined with a JEOL 2010F high resolution transmission electron microscope (HR-TEM) operated at an accelerating voltage of 200 KV, and a JEOL 6335F field emission scanning electron microscope (FE-SEM). The TEM samples were prepared by drop-casting samples dispersed in ethanol onto a carbon-coated holey copper grid followed by drying at room temperature. Photoluminescence (PL) and photoluminescence excitation (PLE) spectra were measured at room temperature using a FP-6500/6600 research-grade fluorescence spectrometer from JASCO with a 150 W Xenon lamp. The samples were prepared by drop-casting onto a quartz slide from a solution of 20 mg calcined nanoparticles in 500 μL of PMMA. In this work, quantum yield was obtained based on the ratio of the intensity of absorption and emission in the drop-casted films.

DISCUSSION

Structural properties of core/multi-shell scintillation nanoparticles

The morphology and particles size of the core-shell structure are shown in Figure 1a for as-prepared mono-dispersed spherical silica core nanoparticles with a diameter of ~220 nm. A thin Gd_2O_3 crust is shown as a clear dark outer region on the light silica core in Figure 1b and 1c, and was confirmed to be Gd_2O_3 with an average 4.5 at% Eu^{3+} concentration by energy dispersed spectrometry (EDS). The selected-area electron diffraction (SAED) pattern shown in the inset in figure 1c indicates that this Gd_2O_3 crust is not well-crystallized before calcination. Figure 1d shows the core-shell nanoparticles after calcining at 800 ℃ for 2 h, and crystalline lattice fringes were observed and confirmed by XRD data. Therefore, by using this sol-gel precipitation method, a uniform shell nanostructure was obtained on the core nanoparticles.

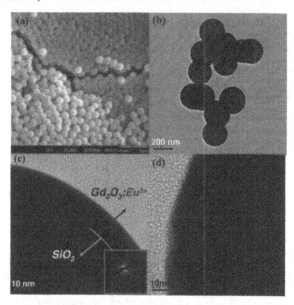

Figure 1. (a) SEM micrograph of mono-dispersed silica nanoparticles synthesized by the Stober method. (b) TEM micrograph of SiO_2/Gd_2O_3:Eu^{3+} nanoparticles before calcination (low magnification). (c) Uncalcined SiO_2/Gd_2O_3:Eu^{3+} nanoparticles with an inset showing selected-area electron diffraction pattern. (d) Nanoparticles of SiO_2/Gd_2O_3:Eu^{3+} calcined at 800℃ 2h.

Figure 2 schematically illustrates the sol-gel precipitation method of core/shell growth. In this method, urea functions as a precipitation agent and the gadolinium and europium metal ions are precipitated on the core material surface followed by further growth at longer reaction time. It is generally believed that the shell thickness is a function of the urea concentration and reaction temperature. As shown in Figure 3, XRD data indicate that the silica core is amorphous, while uncalcined silica/gadolinium oxide core/shell nanoparticles show broad peaks from cubic gadolinium oxide indicating nano-grain size. After the samples were calcined at 800 ℃ for 2 h, a well-crystallized gadolinium oxide cubic phase were obtained (JCPDS: 43-1014).

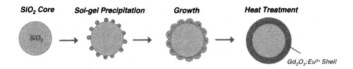

SiO₂ Core   Sol-gel Precipitation   Growth   Heat Treatment

Gd₂O₃:Eu³⁺ Shell

Figure 2. Schematic of the growth of sol-gel core/shell nanoparticles.

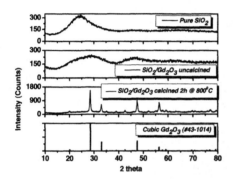

Figure 3. XRD spectra from core/shell nanoparticles.

In order to study the effects of surface passivation on photoluminescence, core/multi-shell scintillation nanoparticles were synthesized, e.g. a SiO₂/Gd₂O₃:Eu³⁺/SiO₂ structure. Previous synthesized SiO₂/Gd₂O₃:Eu³⁺ core/shell nanoparticles were mixed homogeneously (1.8 g) with 60 mL of DI water and 250 mL of ethanol, and followed with addition of 5 mL NH₄OH that was vigorously stirred for 10 min. Then 1 mL of TEOS was added dropwise into the solution and reacted for 3 h. Figure 4a shows a TEM micrograph of uncalcined SiO₂/Gd₂O₃:Eu³⁺/SiO₂ nanoparticles. Three different contrast regions can be seen showing core/multi-shell accordingly as indicated by the arrows. A uniform 10 nm thick silica shell was observed on the doped Gd₂O₃ shell. The sample after calcination at 800 °C for 2 h is shown in Figure 4b, and good coverage by the silica shell is observed.

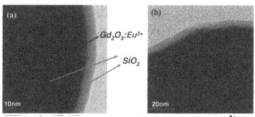

Figure 4. (a) TEM micrograph of SiO₂/Gd₂O₃:Eu³⁺/SiO₂ core/multi-shell nanoparticles before calcination. (b) Calcined SiO₂/Gd₂O₃:Eu³⁺/SiO₂ nanostructure.

176

Optical properties of core/multi-shell scintillation nanoparticles

Photoluminescence properties of core/multi-shell scintillation nanoparticles were investigated by drop-casting 20 mg of nanoparticles in 500 μ L of PMMA onto quartz substrates. Figure 5 shows a schematic diagram of the thin film quantum yield (QY) measurement. The quantum yield can be derived by the ratio of the number of emitted photons to the number of absorbed photons which is defined as S_2 over S_0-S_1, where S_0 is the number of irradiated photons measured by standard blank refractive brick, S1 and S_2 are the number of photons not absorbed and luminous photons emitted by thin film samples, respectively [9]. By measuring these parameters, QY values can be obtained. As shown in Table 1, the QY of $SiO_2/Gd_2O_3:Eu^{3+}$ and $SiO_2/Gd_2O_3:Eu^{3+}/SiO_2$ were 20.3 and 16.5 %, respectively, when excited at 280 nm.

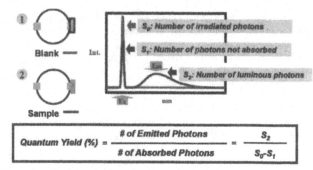

Figure 5. Schematic diagram of thin film quantum yield measurement

Excitation Wavelength (nm)	$SiO_2/Gd_2O_3:Eu^{3+}$ QY (%)	$SiO_2/Gd_2O_3:Eu^{3+}/SiO_2$ QY (%)
280	20.3	16.5

Table 1. Quantum yields of core/multi-shell nanostructure excited at 280 nm.

Figure 6a shows PLE spectra for $SiO_2/Gd_2O_3:Eu^{3+}$ and $SiO_2/Gd_2O_3:Eu^{3+}/SiO_2$. The broad peak at <300 nm is from the oxygen to europium charge transfer band (CTB). The peaks between 350 to 550 nm are the europium self-excitation peaks. It can be seen that for emission at 607 nm, the europium self-excitation peaks are very similar between these two samples. However, the oxygen to europium CTB is much less intense for $SiO_2/Gd_2O_3:Eu^{3+}/SiO_2$ as compared to $SiO_2/Gd_2O_3:Eu^{3+}$ samples. Due to the weak CTB, core/shell/shell samples show lower PL emission as shown in Figure 6b, even though the PL spectra from $^5D_0$ to ^7F_j europium transitions are similar and are dominated by the doublet $^5D_0$-$^7F_2$ transition peaks at 609 and 622nm. The lower PL emission could also be affected by a higher density of interface quenching sites for $SiO_2/Gd_2O_3:Eu^{3+}/SiO_2$.

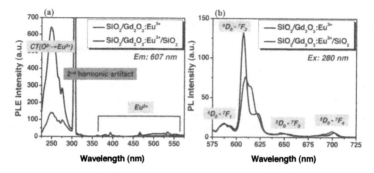

Figure 6. Photoluminescence spectra from $SiO_2/Gd_2O_3:Eu^{3+}$ and $SiO_2/Gd_2O_3:Eu^{3+}/SiO_2$ nanoparticles. (a) PLE spectra for emission at 607 nm. (b) PL spectra excited at 280 nm.

In conclusion, we have demonstrated that core/shell scintillation nanoparticles can be synthesized via a sol-gel precipitation method. The method provides a uniform mono-dispersed spherical core which may be coated by various shell materials and multi-shell schemes. QY measurements showed a value of 20 % for a 13 nm $Gd_2O_3:Eu^{3+}$ shell on a 230 nm SiO_2 core. However, by capping $SiO_2/Gd_2O_3:Eu^{3+}$ with a thin SiO_2 shell, the QY decreased to 16 % which was attributed to less activation of Eu^{3+} through the CTB and/or a higher density of interface quenching sites leading to non-radiative recombination.

ACKNOWLEDGEMENTS

This work was supported by Defense of Threat Reduction Agency (DTRA) Grant No. HDTRA1-08-1-0015. The support of the Major Analytical Instrumentation Center and the Particle Engineering Research Center at the University of Florida are gratefully acknowledged.

REFERENCES

1. F. Wang, Y. Zhang, X. Fan, and M. Wang, *Nanotechnology* **17**, 1527-1532 (2006).
2. D. R. G. J. Daniel Bryan, in *Progress in Inorganic Chemistry*, edited by D. K. Kenneth (2005), p. 47-126.
3. S. C. Erwin, L. Zu, M. I. Haftel, A. L. Efros, T. A. Kennedy, and D. J. Norris, *Nature* **436**, 91 94 (2005).
4. M. Moszynski, *Nuclear Instruments and Methods in Physics Research Section A: Accelerators, Spectrometers, Detectors and Associated Equipment* **505**, 101-110 (2003).
5. M. P. L. L. Peter Reiss, *Small* **5**, 154-168 (2009).
6. D. Gerion, F. Pinaud, S. C. Williams, W. J. Parak, D. Zanchet, S. Weiss, and A. P. Alivisatos, *Journal of Physical Chemistry B* **105**, 8861-8871 (2001).
7. Y. Yang, O. Chen, A. Angerhofer, and Y. C. Cao, *Journal of the American Chemical Society* **130**, 15649-15661 (2008).
8. Stober W, Fink A and Bohn E, *J. Colloid Interface Sci.* **26**, 62 (1968)
9. A. P. Monkman, L.-O. Palsson, *Advanced Materials* **14**, 757-758 (2002).

Mater. Res. Soc. Symp. Proc. Vol. 1164 © 2009 Materials Research Society   1164-L11-04

Prospects for High Energy Resolution Gamma Ray Spectroscopy with Europium-Doped Strontium Iodide

Nerine Cherepy and Stephen A. Payne, Lawrence Livermore National Laboratory, Livermore, CA 94550

Rastgo Hawrami and Arnold Burger, Fisk University, Nashville, TN 37208

Lynn Boatner, Oak Ridge National Laboratory, Oak Ridge, TN 37831

Edgar Van Loef and Kanai Shah, Radiation Monitoring Devices, Watertown, MA 02472

ABSTRACT

Europium-doped strontium iodide scintillators offer a light yield exceeding 100,000 photons/MeV and excellent light yield proportionality, while at the same time, SrI_2 is readily grown in single crystal form. Thus far, our collaboration has demonstrated an energy resolution with strontium iodide of 2.6% at 662 keV and 7.6% at 60 keV, and we have grown single crystals surpassing 30 cm^3 in size (with lower resolution). Our analysis indicates that $SrI_2(Eu)$ has the potential to offer 2% energy resolution at 662 keV with optimized material, optics, and read-out. In particular, improvements in feedstock purity may result in crystal structural and chemical homogeneity, leading to improved light yield uniformity throughout the crystal volume, and consequently, better energy resolution. Uniform, efficient light collection and detection, is also required to achieve the best energy resolution with a $SrI_2(Eu)$ scintillator device.

INTRODUCTION

False alarms in radioisotope identification instrumentation often derive from inadequate energy resolution of the gamma ray spectrometers used. Meanwhile, detection events may be missed altogether if the detectors are too small. For these reasons, we are developing new scintillator materials offering high resolution that are growable in large volumes. Alkaline earth halides are a class of scintillators that appear promising with regards to materials cost, growth, and scintillation yields [1-4]. We have recently found that single crystal divalent Eu-doped Strontium Iodide yields >100,000 photons/MeV in the Eu^{2+} luminescence band (435 nm central wavelength), with a decay time of ~1.2 μs, and excellent light yield proportionality (and hence intrinsic energy resolution) that is superior to that of Ce-doped lanthanum bromide [5-9]. An ongoing coordinated effort between Fisk University, Oak Ridge National Laboratory, Radiation Monitoring Devices, and Lawrence Livermore National Laboratory seeks to improve Strontium Iodide crystal growth and the performance of prototype detectors, with a device goal of 2% resolution at 662 keV and a crystal size of 3" in length by 3" in diameter.

EXPERIMENT

Eu-doped strontium iodide crystals were grown by the vertical Bridgman technique at Fisk University, Radiation Monitoring Devices, and Oak Ridge National Laboratory. Beta-induced luminescence spectra were acquired using a Sr-90/Y-90 (~1 MeV) source, that is expected to be equivalent to that produced by gamma excitation. Gamma ray spectroscopy was performed by optically coupling the crystals with mineral oil to a Hamamatsu R6231-100 PMT (QE at 435 nm of 32%) PMT and wrapping with Teflon reflector materials. The signals from the PMT anode were shaped with a Tennelec TC 244 spectroscopy amplifier (8 μs shaping time) and recorded with an Amptek MCA8000-A multi-channel analyzer; or by interfacing with a Bridgeport Instruments eMorpho digitizer/multi-channel analyzer.

RESULTS AND DISCUSSION

Small crystals (<1 cm^3) of $SrI_2(Eu)$ have been found to provide an energy resolution at 662 kev of 2.5-2.8% [6-9]. Figure 1 shows a small crystal of $SrI_2(Eu)$ offering 662 kev energy resolution comparable to that of $LaBr_3(Ce)$. Figure 2 shows the energy resolution acquired with a Ba-133 source, indicating resolution for $SrI_2(Eu)$ that is markedly superior to that of $LaBr_3(Ce)$ in the lower energy region (Ba-133 main gamma rays are at 81, 276.4, 302.9, 356 and 383.9 kev; some lower energy x-rays appear in the spectrum, as well).

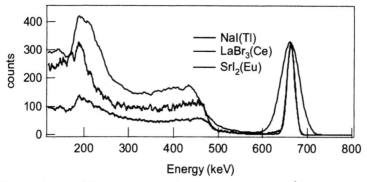

Figure 1. Pulse-height spectra acquired at 662 kev with a ~0.5 cm^3 $SrI_2(5\%Eu)$ crystal grown at Fisk University, along with spectra acquired with a St. Gobain $LaBr_3(Ce)$ crystal and a St. Gobain NaI(Tl) crystal.

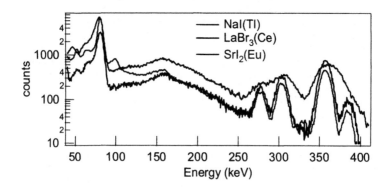

Figure 2. Gamma ray spectra of the same crystals used for Fig. 1, acquired with a ^{133}Ba source.

    A plot of energy resolution as a function of gamma energy shown in Figure 3 indicates that the resolution of SrI$_2$(Eu) is excellent in the low energy region, but is not as good as would be predicted by a simple fit to Poisson statistics at higher gamma energies; an offset, "b" must be added for a good fit to the data. It may be that this energy resolution degradation is due to inhomogeneity of the light yield response, and this is currently being investigated. Assuming the inhomogeneity is due to material non-uniformity, we expect to be able to obtain improved energy resolution, as shown by the blue line (b = 0). The energy resolution at 662 keV for SrI$_2$(Eu) would be 2.3%, if the non-uniformity can be eliminated. Table 1 compares the properties of SrI$_2$(Eu) and LaBr$_3$(Ce).

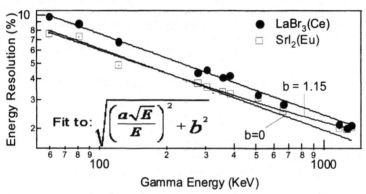

Fit to: $\sqrt{\left(\dfrac{a\sqrt{E}}{E}\right)^2 + b^2}$

Figure 3. The energy resolution as a function of gamma ray energy for SrI$_2$(Eu) and LaBr$_3$(Ce) using Ba-133, Am-241, Co-57, Na-22, Co-60, and Cs-137 sources. Fits to Poisson statistics (plus an offset, b) are shown. For LaBr$_3$(Ce), b = 0, while SrI$_2$(Eu) requires b = 1.15.

Table 1. Comparison of the properties of SrI$_2$(Eu) and LaBr$_3$(Ce) scintillator crystals.

Property	LaBr$_3$(Ce)	SrI$_2$(Eu)	Comparison
Melting Point	783 °C	538 °C	Less thermal stress
Handling	Easily cleaves	Resists cleavage	Better processing
Light Yield	60,000 Ph/MeV	100,000 Ph/MeV	Higher
Radioactivity	La ~ intrinsic backgd	None	Less noise
Gamma absorption (2x3", 662 keV)	22%	24%	Similar
Energy Resolution (662 keV)	2.6% (realized)	2.6% (realized)	Better with uniform crystals
Proportionality contribution	1.6%	1.2%	Favorable
Inhomogeniety contribution	0%	1.2% (current)	Needs development

In addition to the analysis of the data shown in Figure 3, further evidence that the crystal homogeneity should be improved in order to realize the best energy resolution with $SrI_2(Eu)$ is our finding that zone-refined SrI_2 feedstock produces the crystals with the best resolution. Of the 2-5 cm^3 crystals grown by our collaboration, those grown from zone-refined SrI_2 feedstock exhibit the best resolution.

Figure 4 shows how scintillator detectors may be optimized optically by use of a highly reflective Teflon wrapping material from Gore [10]. A ~2 cm^3 $SrI_2(4\%Eu)$ crystal was wrapped, first with St. Gobain Teflon tape, in a tent-like wrapping style, yielding energy resolution of 5.8% at 662 keV. The same crystal was re-wrapped with Teflon tape, tightly around the annulus and top, yielding 3.3% resolution. Finally, the crystal was wrapped with Gore DRP 1 mm material, producing 3.1% resolution at 662 keV. Thus, we find that the absolute reflectivity and uniformity of the reflector wrapping is important in attaining the best energy resolution for a given crystal. We have performed more extensive experiments exploring crystal geometry and wrapping materials that are reported elsewhere [10].

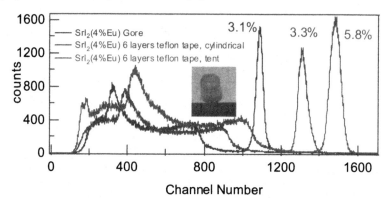

Figure 4. Cs-137 gamma ray (662 keV) pulse height spectra obtained with a $SrI_2(4\%Eu)$ crystal grown at ORNL from feedstock that was zone refined at Fisk University. Signals were collected from the PMT anode with a Bridgeport Instruments eMorpho digitizer. These data indicate that reflectivity and uniformity of wrapping are important considerations for achieving high resolution.

Crystal growth is easiest for cubic crystal structure types, while orthorhombic crystals exhibit modest anisotropy, resulting in crystal growth which is only slightly more subject to orientation dependent differential thermal expansion and growth kinetics than cubic structures. We have found that SrI_2 is not prone to cracking or cleaving, and it is easily ground and polished. So far, the largest single crystal boule (>30 cm^3) has been grown by RMD, shown in Figure 5. The growth of further large crystals and exploration of growth conditions is currently underway.

Figure 5. A ~30 cm^3 single crystal of SrI$_2$(5%Eu), grown by RMD.

CONCLUSION

In summary, we find that strontium iodide crystals doped with europium provide energy resolution comparable to that of cerium doped lanthanum bromide scintillator crystals. The energy-dependent resolution of SrI$_2$(Eu) indicates that it will ultimately surpass the performance of LaBr$_3$(Ce), and is expected to reach 2.3% resolution at 662 keV or better. Single crystal volumes of SrI$_2$(Eu) greater than 30 cm^3 have been grown, thus the scale-up to large homogenous volumes needed for many scintillation counting applications is expected to be achievable.

ACKNOWLEDGMENTS

This work was supported by the Domestic Nuclear Detection Office in the Department of Homeland Security (Alan Janos) and performed under the auspices of the U.S. DOE by Lawrence Livermore National Laboratory under Contract DE-AC52-07NA27344. Oak Ridge National Laboratory is managed for the U.S DOE by UT-Battelle under contract DE-AC05-00OR22725.

REFERENCES

1. R. Hofstadter, E. W. O'Dell, and S. T. Schmidt, *IEEE Trans. Nucl. Sci.* NS-11, p. 12 (1964).
2. R. Hofstadter, "Europium-activated Strontium Iodide Scintillators," US Patent 3,373,279, (1968).
3. N.J. Cherepy, G. Hull, T.R. Niedermayr, A. Drobshoff, S.A. Payne, U.N. Roy, Y. Cui, A. Bhattacharaya, M. Harrison, M. Guo, M. Groza, A. Burger, "Barium iodide single-crystal scintillator detectors," *Proc. SPIE*, vol. 6706, p. 670616 (2007).
4. J. Selling, M.D. Birowosuto, P. Dorenbos, S. Schweizer, "Europium-doped barium halide scintillators for x-ray and γ-ray detections," *J. Appl. Phys.*, vol. 101, p.034901, (2007).
5. N.J. Cherepy, G. Hull, A. Drobshoff, S.A. Payne, E. van Loef, C. Wilson, K. Shah, U.N. Roy, A. Burger, L.A. Boatner, W-S Choong, W.W. Moses "Strontium and Barium Iodide High Light Yield Scintillators," *Appl. Phys. Lett.* vol. 92, p. 083508, (2008).

6. R. Hawrami, M. Groza, Y.Cui, A. Burger, M.D Aggarwal, N. Cherepy and S.A. Payne, "SrI$_2$ a Novel Scintillator Crystal for Nuclear Isotope Identifiers," Proc. SPIE, Vol. 7079, 70790 (2008).

7. CM Wilson, EV Van Loef, J. Glodo, N. Cherepy, G. Hull, SA Payne, W.-S. Choong, WW Moses, KS Shah, "Strontium iodide scintillators for high energy resolution gamma ray spectroscopy," Proc. SPIE, Vol. 7079, 707917, (2008).

8. NJ Cherepy, SA Payne, SJ Asztalos, G Hull, JD Kuntz, T Niedermayr, S Pimputkar, JJ Roberts, RD Sanner, TM Tillotson, E van Loef, CM Wilson, KS Shah, UN Roy, R Hawrami, A Burger, LA Boatner, W.-S. Choong, WW Moses, "Scintillators with Potential to Supersede Lanthanum Bromide," *IEEE Trans. Nucl. Sci.* (in press).

9. E.V.D. van Loef, C.M. Wilson N.J. Cherepy, G. Hull, S.A. Payne, W- S. Choong, W.W. Moses, K.S. Shah, "Crystal Growth and Scintillation Properties of Strontium Iodide Scintillators", *IEEE Trans. Nucl. Sci.* (in press).

10. G. Hull, S. Du, T. Niedermayr, S. Payne, N. Cherepy, A. Drobshoff, and L. Fabris "Light Collection Optimization in Scintillator Based Gamma-Ray Spectrometers," *Nucl. Instr. Meth. A*, vol. 588, p. 384, 2008.

Mater. Res. Soc. Symp. Proc. Vol. 1164 © 2009 Materials Research Society   1164-L11-05

The Synthesis and Structures of Elpasolite Halide Scintillators

Pin Yang[1], F. Patrick Doty[2], Mark A. Rodriguez[1], Margaret R. Sanchez[1], Xiaowong Zhou[2], and Kanai S. Shah[3]
[1]Ceramics and Glass Processing, Sandia National Laboratories,
Albuquerque, NM 87185, U.S.A.
[2]Radiation and Nuclear Detection Materials and Analysis, Sandia National Laboratories,
Livermore, CA 94550, U.S.A.
[3]Radiation Monitoring Devices, Inc.
Watertown, MA 02472, U.S.A.

ABSTRACT

Low-cost, high-performance gamma-ray spectrometers are urgently needed for nonproliferation and homeland security applications. Available scintillation materials fall short of the requirements for energy resolution and sensitivity at room temperature. The emerging lanthanide halide based materials, while having the desired luminosity and proportionality, have proven difficult to produce in the large sizes and low cost required due to highly anisotropic properties caused by the non-cubic crystal structure. New cubic materials, such as the recently discovered elpasolite family (A_2BLnX_6; Ln-lanthanide and X-halogen), hold promise for scintillator materials due to their high light output, proportionality, and toughness. The isotropic nature of the cubic elpasolites leads to minimal thermomechanical stresses during single-crystal solidification, and eliminates the problematic light scattering at the grain boundaries. Therefore, it may be possible to produce these materials in large sizes as either single crystals or transparent ceramics with high production yield and reduced costs. In this study, we investigated the "cubic" elpasolite halide synthesis and studied the structural variations of four different compounds, including $Cs_2NaLaBr_6$, $Cs_2LiLaBr_6$, Cs_2NaLaI_6, and Cs_2LiLaI_6. Attempts to produce a large-area detector by a hot forging technique were explored.

INTRODUCTION

Inorganic scintillators, such as NaI (Tl^{1+}) and CsI(Tl^{1+}), are widely used in radiation detectors at room temperature. These materials play an important role for nuclear and particle physics research, medical imaging, nuclear treaty verification and safeguards, nuclear noprofliferatin monitoring, and geological exploration.[1]

Recent development in the cerium (Ce^{3+}) doped lanthanide halide single crystals, including chlorides[2], bromides[2-4] and iodides[5,6], has shown that these inorganic scintillators exhibit a high light output, a fast decay time, and an outstanding energy resolution, which are excellent for radiation detection applications. However, these single crystals are expensive and difficult to grow in large sizes due to the anisotropic nature of these materials. For example, thermal expansion coefficients for the hexagonal $LaBr_3$ (Space group: P63/m) along its c axis and normal to the prismatic plane are 13.46 $X10^{-6}/°C$ and 28.12 X $10^{-6}/°C$, respectively.[7] The difference can create large thermomechanical stresses in the crystal during solidification process. Furthermore, these materials have extremely limited ductility and low fracture toughness in comparison to traditional halide salts. Cracks can be easily initiated and propagating along their prismatic planes.[7] These factors limit the available crystal sizes, increase manufacturing costs, and hamper the wide spread of these materials for radiation detection applications.

An alternative ceramic approach was explored recently to address these manufacturability issues. For this approach, lanthanide halide powder (with Ce^{3+} activator) was first pressed to ~50% of its theoretical density, and subsequently sintered at higher temperature to produce a dense body. It was quickly learned from the thermal gravimetric analysis measurement that the volatility of bromine on the surface of the lanthanide bromide powder can be serious above 330 °C even though the bulk material such as a single crystal remains thermally stable at these temperatures.[8] This issue pose a challenge for densification of ceramics where an effective densification for these ceramic compacts begins above 600°C as determined by the sintering shrinkage curve.[8] The difference between sintering temperature and temperature where material becomes unstable can cause a significant weight loss and deviation from stoichometry[9] during densification. In fact, sintering these fine $LaBr_3$ powder compacts under vacuum will produce darkened samples[8] as $LaBr_3$ is reduced to La_2Br_5.

High Z, no self-activity, and thermally stable cubic halides are desirable for single crystal growth or ceramic fabrication. An isotropic cubic material will minimize the thermomechanical stresses during single crystal growth and eliminate problematic light scattering at the grain boundaries of a polycrystalline ceramic body. Therefore, high symmetry enables large size single crystals or transparent ceramics to be produced with high production yield and reduced costs. Among more than 500 lanthanide halide compounds in the literature,[10] the elpasolite family was chosen, as more than 80% of these compounds possess a cubic structure (space group Fm-3m). Four high Z, lanthanum-based elpasolite halides, $Cs_2NaLaBr_6$, $Cs_2LiLaBr_6$, Cs_2NaLaI_6, and Cs_2LiLaI_6 were initially selected for the synthesis and characterization due to their interesting scintillation properties (Table I). Special emphasis was placed on the material synthesis, structure characterization, and thermal stability evaluation with respect to processing conditions.

Table I. The performance of Ce^{+3} doped elpasolite halide scintillators (RMD data)

Material	Density (g/cm^3)	Light output (Photons/MeV)	Decay time (ns)	Non-proportionality (60-1275 keV)
Cs_2NaLaI_6[11]	~ 4.2	54,000 – 60,000	50	< 1
Cs_2LiLaI_6	~ 4.3	> 50,000	51	1.6
$Cs_2LiLaBr_6$	~ 4.1	50,000 – 55,000	65	1.7
$Cs_2NaLaBr_6$	~ 3.9	12,000	55	2.9

EXPERIMENTAL PROCEDURE

Three different approaches, including solution synthesis,[12,13] solid-state reaction, and melt synthesis, were explored to fabricate the Ce^{3+} (5 mol %) doped lanthanum-based elpasolite halides. Initial attempts to synthesize these compounds by salt-acid[12] and metal-acid[13] reactions followed by vacuum dehydration were unsuccessful. X ray diffraction and thermal analysis results indicated that these synthesized powders were a mixture of salts, presumably due to a large difference in solubility of these salts in the acid solution. As a result, halide salts, including CsI, NaI, LiI, LaI$_3$, CeI$_3$, CsBr, NaBr, LiBr, LaBr$_3$, and CeBr$_3$, were used to directly synthesize these elpasolites. For these processes, high purity anhydrous bromide and iodide salts (>

99.999%; except LaI$_3$ which is 99.9%) from Alfa-Aldrich were weighed according to their chemical formulations. These salts were mixed and ground with a mortar and a pestle for 2 hours. These halide salts, in particular the LaBr$_3$ and the LaI$_3$, were found to be extremely oxophilic. They tend to reduce the refractory materials in the box furnace and form oxyhalides (LaOBr and LaOI) during solid-state reactions. To prevent such a reaction, the ground, mixed salts were loaded into quartz ampoule in an argon filled glovebox and subsequently vacuum sealed for the melting synthesis. Lanthanum-based elpasolite halides were finally successfully synthesized by melting and solidification of these halide salts in the vacuum sealed quartz ampoules. Experimental results indicated that adequate mixing and controlling the reaction kinetics are critical for achieving phase-pure elpasolites (see next section for details). In addition, small single crystals of lanthanum-based elpasolite halides produced by small-scale growth methods from Radition Monitoring Devices, Inc. (Watertown, MA) have been provided for initial crystal structure determination.

RESULTS AND DISCUSSION

Melting synthesis and crystal structure

Preliminary experiments showed that phase-pure elapsolite halides could not be produced by a quick melting and solidification of the ground alkaline and lanthanum halide salts. Differential scanning calorimetric data indicated that there were some residual, unreacted halide salts in the melt as evidenced by several thermal events observed during the solidification process. The presence of these salts was later confirmed by X-ray diffraction analysis. However, these events progressively disappeared as the material went through thermal cycles, while the exothermic peak corresponding to the solidification of elpasolite halide increased at the expense of these residual unreacted phases. These observations are illustrated in Figure 1 for the Cs$_2$NaLaI$_6$, where the dotted, dashed, and solid lines represent the first, the second, and the third cooling cycles, respectively. Close examination of these thermal events, one will find none of these temperatures correspond to the melting points of the starting materials (i.e., LaI$_3$ – 778°C, NaI – 654°C, and CsI – 624°C).; instead they are close to the eutectic points in the three constituent binary systems for the CsI-NaI-LaI$_3$ ternary system. For example, major thermal events found at 543°C, 454°C and 429°C are well matched to the eutectic points found in Cs$_3$LaI$_6$-CsI (539°C),[14] NaI-LaI$_3$(454°C),[15] and CsI-NaI (426°C)[15] binary phase diagrams. However, the dominate thermal event (the highest peak) observed in the cooling cycle is the solidification of the elpasolite phase near 540° C. This peak becomes sharper as the number of thermal cycle increases. Finally, the melting point of Cs$_2$NaLaI$_6$ was well defined at 540°C with no-detectable foreign phases present on the solidification curve. The merging and disappearing of some of exothermal peaks on Figure 1 depict the reaction kinetics involved with the Cs$_2$NaLaI$_6$ elpasolite formation, as the chemical reaction approaches completion through the subsequent thermal cycles. These observations immediately suggestion that reaction kinetics is important for the formation of elpasolites and a longer reaction time or higher temperature is required to complete the reaction.

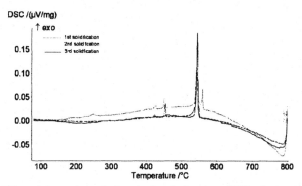

Figure 1. The differential scanning calorimetric data during the solidification cycles for the Cs_2NaLaI_6. Data were collected at 3°C/min under a flowing argon condition.

Several initial attempts to form a phase-pure Cs_2NaLaI_6 from the melt, based on the aforementioned observations were unsuccessful. X-ray diffraction data indicated these samples still had traces of foreign phases after homogenization at 30°C above the highest melting point of the salts ($LaBr_3$ at 788°C or LaI_3 at 778°C) for 10 hours. It was found that agitating the melt (by shaking the ampoule in the furnace) at a higher temperature helped the elpasolite phase formation, presumably due to the density difference between these molten salts. To minimize this gravitational separation, a two-zone, vertical furnace was constructed and configured with a thermal gradient of about 50°C per inch from bottom (hot-end) to top (cold-end) of the ampoule to provide additional convection mixing. Phase-pure solidified Cs_2NaLaI_6 ingot, as determined by the powder X ray diffraction data, was obtained after melting at 850 °C (the maximum temperature at the bottom of the ampoule) for 10 hours. After melting, the furnace was slowly cooled to room temperature. Other compounds were made by the same thermal profile, without any attempt to optimization the process.

Structural refinements of these elpasolites were performed on our samples and crushed single crystals obtained from Radiation Monitoring Devices Inc. Lattice parameters of these single crystals are reported in Table II. A comparison of the X-ray diffraction data and theoretically calculated intensity for $Cs_2LiLaBr_6$ is given in Figure 2. Results show that there were minor intensity mismatches (at ~31.5° and ~45.5°) and the diffraction pattern indexed well with a prototype cubic elpasolite structure[16] (Fm-3m). Therefore, $Cs_2LiLaBr_6$ belongs to the same space group as other cubic alkaline halide scintillators such as NaI and CsI. In addition, the Goldschmidt tolerance factor[17] (t, see insert in Figure 2 for the formulation and corresponding structure) for elpasolites was calculated based on the ionic radii and the coordination number of cations and anions in the lattice.[18] The tolerance factor is a geometric factor that gives a necessary but not sufficient condition for the formation and lattice distortion of perovskite-type complex halides, based on a hard-sphere model. This factor provides a first-order estimation in terms of possible lattice distortion for a cubic cell, which is closely related to the degree of anisotropy in the lattice. Implications can be drawn for material selection in order to minimize the anisotropic thermomechanical stresses developed in single crystal growth or to reduce the amount of light scattering in polycrystalline ceramics.

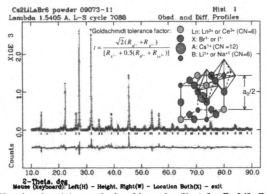

Cs2LiLaBr6 powder 09073-11   Hist 1
Lambda 1.5405 A. L–S cycle 7088   Obsd. and Diff. Profiles

Figure 2. X-ray diffraction data (+) vs. calculated intensity (line) for $Cs_2LiLaBr_6$. Inserts include the cubic elpasolite structure and the Goldschmidt tolerance factor.

Table II. Crystal structure and tolerance factor for lanthanum-based elpasolite halides.

Material	Crystal structure	Lattice parameters (Å)			Tolerance factor (t)
		a	b	c	
$Cs_2LiLaBr_6$	Cubic (Fm-3m)	11.2890	11.2890	11.2890	0.950
Cs_2LiLaI_6					0.935
$Cs_2NaLaBr_6$[10]	Tetragonal	11.620	11.620	11.605	0.909
$Cs_2NaLaI6$	Orthorhombic (Pnma)	8.7615	12.4361	8.6270	0.897

Note, as the tolerance factor deviates from unity, these elpasolites change from a cubic ($Cs_2LiLaBr_6$) to a tetragonal ($Cs_2NaLaBr_6$) finally to an orthorhombic structure (Cs_2NaLaI_6). This result demonstrates that the calculated tolerance factor can be used for screening material candidates in the large eplasolite family (> 240 compounds available)[10], which will reduce the lattice distortion and the thermomechanical stress for single crystal growth and minimize the light scattering in polycrystalline ceramics.

Thermal stability and hot forging

Figure 3 shows the thermal gravimetric analysis (TGA) and differential thermal analysis (DTA) results for $Cs_2LiLaBr_6$. Data were collected with a heating and a cooling rate of 3°C/min under flowing argon condition. The only thermal event observed during the heating cycle corresponds to the melting point of $Cs_2LiLaBr_6$ (T_m = 478°C), as indicated by the endothermic peak on DTA curve. Note during the heating cycle, the material started to lose mass above 460 °C, and continued losing mass during the holding period (i.e., 15 minutes at 820°C). Before the cooling cycle, the total weight loss had exceeded 75 wt% and was out of the detection range for the TGA unit. As a result, during cooling the solidification event was not detected. This observation was consistent with an empty alumina pan left in the thermal analysis unit. Therefore, we define the onset temperature for the weight loss on the TGA curve as T_u where material becomes thermally unstable. This temperature sets an upper limit for the densification of $Cs_2LiLaBr_6$ ceramic compacts in order to prevent significant weight loss. A summary for the melting points (T_m) and T_u for these elpasolite halides is given in Table IV.

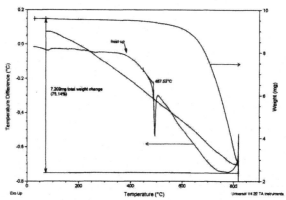

Figure 3. The melting point (~488°C) and the weight loss for $Cs_2LiLaBr_6$ during a heating cycle, as determined by the TGA (on the left) and the TGA (on the right).

Table IV. Summary of T_m and T_u for lanthanum-based elpasolite halides (5 mole% Ce^{3+}).

Material	T_m (°C)	T_u (°C)
Cs_2NaLaI_6	540	> 420
$Cs_2NaLaBr_6$	595	> 510
Cs_2LiLaI_6	476	> 350
$Cs_2LiLaBr_6$	488	> 460

$Cs_2LiLaBr_6$ was chosen for the first hot forging experiment because it has a cubic structure among all these elpasolite halides. A preliminary hot forging experiment was conducted for $Cs_2LiLaBr_6$ at 478°C for 4 hours, which was 10°C below its T_m and 17°C above T_u. Therefore, it is important to assure that full density can be achieved within a short time span by the hot forging process. A solidified $Cs_2LiLaBr_6$ ingot with a diameter of 0.98 cm and thickness of 0.7 mm (see Figure 4 (a)) was hot forged at 478°C with a pressure ramping rate at 100 psi/min. The polycrystalline ingot was centered in a 2.54 cm diameter graphite die with graphite foil liner and spacers to separate the sample in direct contact with the graphite die assembly. It was found that the cylinder specimen was completely deformed into a thin disk (2.54 cm diameter, 0.104 cm thick) and the densification was completed within the first 5 minutes. The forged sample is shown in Figure 4 (b) where graphite foil was still attached to the sample; therefore, the transparency of the material could not be determined. Attempts to remove the foil from the thin disk by a razor blade were unsuccessful, and the sample broke into pieces. However, this experiment demonstrates that the hot forging technique can offer a short processing time and an effective densification for these elpasolite halides. These factors are important to remove pores in ceramic body and produce highly transparent ceramics without causing significant weight loss or deviation from its stoichometry. The ability to sustain such a large plastic deformation for the $Cs_2LiLaBr_6$ can be attributed to its relatively soft heavy ionic

190

nature, as well as many active slip systems available in the face-center-cubic structure. This experiment demonstrates the feasibility of making large-area, transparent detectors by a hot forging technique similar to these reported for NaI and CsI. We will continue pursuing this experiment with large size samples, identify suitable spacer material, and expand and refine our experiments for other elpasolite halides.

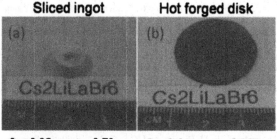

Sliced ingot   **Hot forged disk**

$\Phi = 0.98$ cm; $t = 0.70$ cm $\Phi = 2.54$ cm; $t = 0.104$cm

Figure 4. The change of sample geometry for a sliced $Cs_2LiLaBr_6$ ingot before (a) and after (b) hot forged. The aspect ratio (D/t) of the sample changed from 1.4 to 24.4 after hot forging.

CONCLUSIONS

A hard-sphere model, based on Goldschmidt tolerance factor, has shown to give a good predication on lattice distortion for all the elpasolite halides studied in this work. This simple parameter is useful in selection of new compounds from a large elpasolite halide family for crystal growth or fabrication of transparent ceramics for scintillation applications. Results show that as the tolerance factor deviates from unity, these elpasolite halides change their symmetry from cubic to tetragonal, and finally to orthorhombic. In addition, our preliminary hot forging experiment demonstrated the feasibility of achieving a fully dense elpasolite polycrystalline ceramic without deviating from its stoichometry. Implications are important for making low-cost, large, transparent polycrystalline detectors.

ACKNOWLEDGMENTS

The authors would like thank Denise N. Bencoe and Clay S. Newton for their assistance in thermal analysis and hot forging experiment. The financial support from DoE NA-22 Advanced Material Portfolio is also greatly appreciated. Sandia is a multiprogram laboratory operated by Sandia Corporation, a Lockheed Martin Company, for the United States Department of Energy's National Nuclear Security Administration under contract DE-AC04-94AL85000.

REFERENCES

1. G. Knoll, Radiation Detection and Measurement, 3rd ed., New York: Wiley, 1999.
2. O. Guillot-Noël, J. T. M. de Hass, P. Dorenbos, C.W.E. van Eijk, K. Krämer, and H. U. Güdel, "Optical and scintillation properties of cerium-doped LaCl$_3$, LuBr$_3$, and LuCl$_3$," *J. Lumin.*, **85** 21-35 (1999).

3. E.V.D. van Loef, P. Dorenbos, C. W. E. van Eijk, K. Krämer, and H. U. Gűdel, "High energy resolution scintillator: Ce^{+3} activated LaBr$_3$," *Appl. Phys. Lett.*, **79** [10] 1573-1575 (2001).

4. K. S. Shah, J. Glodo, W. H. Higgins, E. V. D. van Loef, W. W. Moses, S. E. Derenzo and M. J. Weber, "CeBr$_3$ scintillators for gamma-ray spectroscopy," *IEEE Trans. Nucl. Sci.*, 52 [6] 3157-3159 (2005).

5. M. D. Birowosuto, P. Dorenbos, and C. W. E. van Eijk, K.W. Krämer, and H. U. Gűdel, "High-light-output scintillator for photodiode readout: LuI$_3$: Ce^{3+}," *J. Appl. Phys.*, **99** 123520 (2006).

6. J. Glodo, W. M. Higgins, E. V. D. van Loef, and K. S. Shah, "GdI$_3$:Ce – A new gamma and neutron scintillator," *IEEE Nucl. Sci. Symposium Conference Record*, 1574-1577 (2006).

7. F. P. Doty, D. McGregor, M. Harrison, K. findley and R Polichar, "Structure and property of lanthanide halides," SPIE **6707**, 670705 (2007).

8. P. Yang, T. J. Boyle, N. S. Bell, M. R. Sanchez, L. A. M. Ottley, and C. F. Chen, "Fabrication of large-volume, low-cost ceramic lanthanum halide scintillators for gamma ray detection," Sandia Report, Sandia National Laboratories, SAND2008-6978 and SAND2007-0719.

9. K. Krämer, T. Schleid, M. Schulze, W. Urland, and G. Mayer, "Three Bromides of Lanthanum: LaBr$_2$, La$_2$Br$_5$ and LaBr$_3$," *Z. Anorg. Allg. Chem.*, **575** 61-70 (1989).

10. G. Meyer, "The synthesis and structures of complex rare-earth halides," *Prog. Solid St. Chem.*, **14**, 141-219 (1982).

11. J. Glodo, E. V. D. van Loef, W. M. Higgins, and K. S. Shah, "Scinillation Properties of Cs$_2$NaLaI$_6$:Ce," 2006 IEEE Nuclear Science Symposium Conference Record N30-164 (2006).

12. M. E. Villafuerte-Catrejon, M. R. Estrada, J. Gomez-Lara, J. Duque, and R. Pomes, " Crystal structure of Cs$_2$KTbCl$_6$ and Cs$_2$KEuCl$_6$ by powder X-ray diffraction," *J. Solid State Chem.*, **132** 1-5 (1997).

13. T. J. Boyle, P. Yang, L. A. M. Ottley, M. A. Rodriguez, T. M. Alam and S. Hoppe, "Synthesis, characterization, and processing of hydrates and anhydrous species of simple and mixtures lanthanum halide materials for scintillator application," in preparation (2008).

14. J. Kutscher and A. Schneide, "Chemistry of rare earths in melten alkaline halides. 8. Study on diagrams of state of lanthanide (III) iodides in mixture of alkaline iodides," *Z. Anorg. Allg. Chem.*, **386** [1] 38-46 (1971).

15 J. Sangster and A. D. Pelton, "Phase diagrams and thermodynamic properties of the 70 binary alkali-halide systems having common ions, " *J. Phys. Chem.*, **16** [3] 509-561 (1987).

16. C. Reber, H. Gűdel, G. Meyer, T. Schleid, C. A. Dual, "Optical spectroscopic and structural-properties of V^{+3}-doped fluoride, chloride, and bromide elpasolite lattices," *Inorganic Chem.*, **28** [16] 3249-3581 (1989).

17. L. Liu, W. Lu, and N. Chen, "On the criteria of formation and lattice distortion of perovskite-type complex halides," *J. Phys. Chem. Solids*, **65** 855-860 (2004).

18. R. D. Shannon, "Revised effective ionic radii and systematic studies of interatomic distances in halides and chalogenides," *Acta Cryst.*, **A32** 751-766 (1976)

AUTHOR INDEX

SUBJECT INDEX

195

Printed in the United States
by Baker & Taylor Publisher Services